STUDENT SOLUTIONS MANUAL
VOLUME ONE: CHAPTERS 1–16

SEARS & ZEMANSKY'S

COLLEGE PHYSICS

8TH EDITION

YOUNG & GELLER

A. Lewis Ford
Texas A & M University

PEARSON

Addison
Wesley

San Francisco   Boston   New York
Cape Town   Hong Kong   London   Madrid   Mexico City
Montreal   Munich   Paris   Singapore   Sydney   Tokyo   Toronto

Editor-in-Chief:	Adam Black
Associate Editor:	Chandrika Madhavan
Managing Editor, Production:	Corinne Benson
Production Supervisor:	Jane Brundage
Executive Marketing Manager:	Christy Lawrence
Manufacturing Buyer:	Pam Augspurger
Production Services:	WestWords, Inc.
Cover Design:	Seventeenth Street Studios

Cover Photo Credit: © Ken Findlay/Dorling Kindersley

ISBN 0-8053-9370-6

PEARSON
Addison
Wesley

1 2 3 4 5 6 7 8—CRS—10 09 08 07 06

www.aw-bc.com

CONTENTS

PREFACE

This *Student Solution Manual, Chapters 1-16* covers chapters 1 through 16 of *College Physics* 8th edition by Young and Geller. It contains detailed solutions for approximately one third of the end-of-chapter Problems. The Problems included in this manual are selected solely from the odd-numbered Problems in the text (for which the answers are tabulated in the back of the textbook). The Problems included were not selected at random but rather were carefully chosen to include at least one representative example of each problem type. The remaining Problems, for which solutions are not given here, constitute an ample set of problems for students to tackle on their own.

All solutions are done in the Set Up/Solve/Reflect framework used in the textbook. In most cases rounding was done in intermediate steps, so you may obtain slightly different results if you handle the rounding differently.

This manual greatly expands the set of worked-out examples that accompany the presentation of physics laws and concepts in the text. This manual was written to provide students with models to follow in working physics problems. The problems are worked out in the manner and style in which students should carry out their own problem solutions.

We have made every effort to be accurate and correct in the solutions, but if you find errors or ambiguities it would be very helpful if you would point these out to the author.

Lewis Ford
Physics Department MS 4242
Texas A&M University
College Station, TX 77843-4242
ford@physics.tamu.edu

MODELS, MEASUREMENTS, AND VECTORS

Problems 1, 5, 9, 15, 19, 23, 27, 29, 33, 35, 37, 41, 45, 47, 49, 51, 53, 59, 61, 65

Solutions to Problems

1.1. Set Up: We know the following equalities: $1 \text{ mg} = 10^{-3} \text{ g}$; $1 \, \mu\text{g} = 10^{-6} \text{ g}$; $1 \text{ kilohms} = 1000 \text{ ohms}$; and $1 \text{ milliamp} = 10^{-3} \text{ amp}$.
Solve: In each case multiply the quantity to be converted by unity, expressed in different units. Construct an expression for unity so that the units to be changed cancel and we are left with the new desired units.

(a) $(2400 \text{ mg/day}) \left(\dfrac{10^{-3} \text{ g}}{1 \text{ mg}} \right) = 2.40 \text{ g/day}$

(b) $(120 \, \mu\text{g/day}) \left(\dfrac{10^{-6} \text{ g}}{1 \, \mu\text{g}} \right) = 1.20 \times 10^{-4} \text{ g/day}$

(c) $(500 \text{ mg/day}) \left(\dfrac{10^{-3} \text{ g}}{1 \text{ mg}} \right) = 0.500 \text{ g/day}$

(d) $(1500 \text{ ohms}) \left(\dfrac{1 \text{ kilohm}}{10^{3} \text{ ohms}} \right) = 1.50 \text{ kilohms}$

(e) $(0.020 \text{ amp}) = \left(\dfrac{1 \text{ milliamp}}{10^{-3} \text{ amp}} \right) = 20 \text{ milliamps}$

Reflect: In each case, the number representing the quantity is larger when expressed in the smaller unit. For example, it takes more milligrams to express a mass than to express the mass in grams.

1.5. Set Up: We need to apply the following conversion equalities: $1000 \text{ g} = 1.00 \text{ kg}$, $100 \text{ cm} = 1.00 \text{ m}$, and $1.00 \, L = 1000 \text{ cm}^3$.

Solve: (a) $(1.00 \text{ g/cm}^3) \left(\dfrac{1.00 \text{ kg}}{1000 \text{ g}} \right) \left(\dfrac{100 \text{ cm}}{1.00 \text{ m}} \right)^3 = 1000 \text{ kg/m}^3$

(b) $(1050 \text{ kg/m}^3) \left(\dfrac{1000 \text{ g}}{1.00 \text{ kg}} \right) \left(\dfrac{1.00 \text{ m}}{1000 \text{ cm}} \right)^3 = 1.05 \text{ g/cm}^3$

(c) $(1.00 \, L) \left(\dfrac{1000 \text{ cm}^3}{1.00 \, L} \right) \left(\dfrac{1.00 \text{ g}}{1.00 \text{ cm}^3} \right) \left(\dfrac{1.00 \text{ kg}}{1000 \text{ g}} \right) = 1.00 \text{ kg}$; $(1.00 \text{ kg}) \left(\dfrac{2.205 \text{ lb}}{1.00 \text{ kg}} \right) = 2.20 \text{ lb}$

Reflect: We could express the density of water as 1.00 kg/L.

1.9. Set Up: We apply the equalities of $1 \text{ km} = 0.6214 \text{ mi}$ and $1 \text{ gal} = 3.788 \, L$.

Solve: $(37.5 \text{ mi/gal}) \left(\dfrac{1 \text{ km}}{0.6214 \text{ mi}} \right) \left(\dfrac{1 \text{ gal}}{3.788 \, L} \right) = 15.9 \text{ km/L}$

Reflect: Note how the unit conversion strategy, of cancellation of units, automatically tells us whether to multiply or divide by the conversion factor.

1.15. Set Up: In each case, round the last significant figure.
Solve: (a) 3.14, 3.1416, 3.1415927 **(b)** 2.72, 2.7183, 2.7182818 **(c)** 3.61, 3.6056, 3.6055513
Reflect: All of these representations of the quantities are imprecise, but become more precise as additional significant figures are retained.

1.19. Set Up: We are given the relation density $=$ mass/volume $= m/V$ where $V = \frac{4}{3}\pi r^3$ for a sphere. From Appendix F, the earth has mass of $m = 5.97 \times 10^{24}$ kg and a radius of $r = 6.38 \times 10^6$ m whereas for the sun at the end of its lifetime, $m = 1.99 \times 10^{30}$ kg and $r = 7500$ km $= 7.5 \times 10^6$ m. The star possesses a radius of $r = 10$ km $= 1.0 \times 10^4$ m and a mass of $m = 1.99 \times 10^{30}$ kg.
Solve: (a) The earth has volume $V = \frac{4}{3}\pi r^3 = \frac{4}{3}\pi(6.38 \times 10^6 \text{ m})^3 = 1.088 \times 10^{21}$ m^3.

$$\text{density} = \frac{m}{V} = \frac{5.97 \times 10^{24} \text{ kg}}{1.088 \times 10^{21} \text{ m}^3} = (5.49 \times 10^3 \text{ kg/m}^3)\left(\frac{10^3 \text{ g}}{1 \text{ kg}}\right)\left(\frac{1 \text{ m}}{10^2 \text{ cm}}\right)^3 = 5.49 \text{ g/cm}^3$$

(b) $V = \frac{4}{3}\pi r^3 = \frac{4}{3}\pi(7.5 \times 10^6 \text{ m})^3 = 1.77 \times 10^{21}$ m^3

$$\text{density} = \frac{m}{V} = \frac{1.99 \times 10^{30} \text{ kg}}{1.77 \times 10^{21} \text{ m}^3} = (1.1 \times 10^9 \text{ kg/m}^3)\left(\frac{1 \text{ g/cm}^3}{1000 \text{ kg/m}^3}\right) = 1.1 \times 10^6 \text{ g/cm}^3$$

(c) $V = \frac{4}{3}\pi r^3 = \frac{4}{3}\pi(1.0 \times 10^4 \text{ m})^3 = 4.19 \times 10^{12}$ m^3

$$\text{density} = \frac{m}{V} = \frac{1.99 \times 10^{30} \text{ kg}}{4.19 \times 10^{12} \text{ m}^3} = (4.7 \times 10^{17} \text{ kg/m}^3)\left(\frac{1 \text{ g/cm}^3}{1000 \text{ kg/m}^3}\right) = 4.7 \times 10^{14} \text{ g/cm}^3$$

Reflect: For a fixed mass, the density scales as $1/r^3$. Thus, the answer to (c) can also be obtained from (b) as

$$(1.1 \times 10^6 \text{ g/cm}^3)\left(\frac{7.50 \times 10^6 \text{ m}}{1.0 \times 10^4 \text{ m}}\right)^3 = 4.7 \times 10^{14} \text{ g/cm}^3.$$

1.23. Set Up: The mass can be calculated as the product of the density and volume. The volume of the washer is the volume V_d of a solid disk of radius r_d minus the volume V_h of the disk-shaped hole of radius r_h. In general, the volume of a disk of radius r and thickness t is $\pi r^2 t$. We also need to apply the unit conversions 1 m^3 $= 10^6$ cm^3 and 1 g/cm^3 $= 10^3$ kg/m^3.
Solve: The volume of the washer is:

$$V = V_d - V_h = \pi(r_d^2 - r_h^2)t = \pi[(2.25 \text{ cm})^2 - (0.625 \text{ cm})^2](0.150 \text{ cm}) = 2.20 \text{ cm}^3$$

The density of the washer material is 8600 kg/m$^3[(1 \text{ g/cm}^3)/(10^3 \text{ kg/m}^3)] = 8.60$ g/cm^3. Finally, the mass of the washer is: mass $= (\text{density})(\text{volume}) = (8.60 \text{ g/cm}^3)(2.20 \text{ cm}^3) = 18.9$ g.
Reflect: This mass corresponds to a weight of about 0.7 oz, a reasonable value for a washer.

1.27. Set Up: The number of kernels can be calculated as $N = V_{\text{bottle}}/V_{\text{kernel}}$. Based on an Internet search, Iowan corn farmers use a sieve having a hole size of 0.3125 in. $\cong$ 8 mm to remove kernel fragments. Therefore estimate the average kernel length as 10 mm, the width as 6 mm and the depth as 3 mm. We must also apply the conversion factors 1 L $= 1000$ cm^3 and 1 cm $= 10$ mm.
Solve: The volume of the kernel is: $V_{\text{kernel}} = (10 \text{ mm})(6 \text{ mm})(3 \text{ mm}) = 180$ mm^3. The bottle's volume is: $V_{\text{bottle}} = (2.0 \text{ L})[(1000 \text{ cm}^3)/(1.0 \text{ L})][(10 \text{ mm})^3/(1.0 \text{ cm})^3] = 2.0 \times 10^6$ mm^3. The number of kernels is then $N_{\text{kernels}} = V_{\text{bottle}}/V_{\text{kernels}} \approx (2.0 \times 10^6 \text{ mm}^3)/(180 \text{ mm}^3) = 11{,}000$ kernels.
Reflect: This estimate is highly dependent upon your estimate of the kernel dimensions. And since these dimensions vary amongst the different available types of corn, acceptable answers could range from 6,500 to 20,000.

1.29. Set Up: Estimate the thickness of a dollar bill by measuring a short stack, say ten, and dividing the measurement by the total number of bills. I obtain a thickness of roughly 1 mm. From Appendix F, the distance from the earth to the moon is 3.8×10^8 m. The number of bills is simply this distance divided by the thickness of one bill.
Solve: $N_{\text{bills}} = \left(\frac{3.8 \times 10^8 \text{ m}}{0.1 \text{ mm/bill}}\right)\left(\frac{10^3 \text{ mm}}{1 \text{ m}}\right) = 3.8 \times 10^{12}$ bills $\approx 4 \times 10^{12}$ bills

Reflect: This answer represents 4 trillion dollars! The cost of a single space shuttle mission in 2005 is significantly less—roughly 1 billion dollars.

1.33. Set Up: The cost would equal the number of dollar bills required; the surface area of the U.S. divided by the surface area of a single dollar bill. By drawing a rectangle on a map of the U.S., the approximate area is 2600 mi by 1300 mi or 3,380,000 mi^2. This estimate is within 10 percent of the actual area, 3,794,083 mi^2. The population is roughly 3.0×10^8 while the area of a dollar bill, as measured with a ruler, is approximately $6\frac{1}{8}$ in. by $2\frac{5}{8}$ in.
Solve: $A_{\text{U.S.}} = (3,380,000 \text{ mi}^2)[(5280 \text{ ft})/(1 \text{ mi})]^2[(12 \text{ in.})/(1 \text{ ft})]^2 = 1.4 \times 10^{16} \text{ in.}^2$

$$A_{\text{bill}} = (6.125 \text{ in.})(2.625 \text{ in.}) = 16.1 \text{ in.}^2$$

$$\text{Total cost} = N_{\text{bills}} = A_{\text{U.S.}}/A_{\text{bill}} = (1.4 \times 10^{16} \text{ in.}^2)/(16.1 \text{ in.}^2/\text{bill}) = 9 \times 10^{14} \text{ bills}$$

$$\text{Cost per person} = (9 \times 10^{14} \text{ dollars})/(3.0 \times 10^8 \text{ persons}) = 3 \times 10^6 \text{ dollars/person}$$

Reflect: The actual cost would be somewhat larger, because the land isn't flat.

1.35. Set Up: The displacement vector $\vec{d}$ is directed from the initial position of the object to the final position. In each case, its magnitude d is the length of the line that connects points 1 and 2.
Solve: **(a)** The initial position (point 1), the final position (point 2), the displacement vector $\vec{d}$ and its direction ϕ is shown in Figure 1.35 for each case.

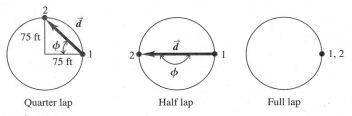

Figure 1.35

(b) *Quarter lap:* The magnitude is $d = \sqrt{r^2 + r^2} = \sqrt{2}r = \sqrt{2}(75 \text{ ft}) = 106 \text{ ft}$. From Figure 1.35, $\tan\phi = r/r = 1.0$ or $\phi = 45°$. *Half lap:* The magnitude is $d = 2r = 150 \text{ ft}$. The direction is $\phi = 180°$. *Full lap:* The final position equals the initial position. The displacement is therefore equal to zero and the direction is undefined.
Reflect: In each case, the magnitude of the displacement is less than the distance traveled.

1.37. Set Up: Draw the vectors to scale on graph paper, using the tip to tail addition method. For part (a), simply draw $\vec{B}$ so that its tail lies at the tip of $\vec{A}$. Then draw the vector $\vec{R}$ from the tail of $\vec{A}$ to the tip of $\vec{B}$. For (b), add $-\vec{B}$ to $\vec{A}$ by drawing $-\vec{B}$ in the opposite direction to $\vec{B}$. For (c), add $-\vec{A}$ to $-\vec{B}$. For (d), add $-\vec{A}$ to $\vec{B}$.
Solve: The vector sums and differences are shown in Figure. 1.37a-d.

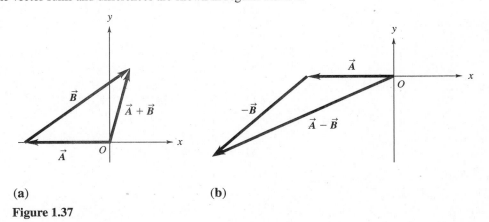

(a)

(b)

Figure 1.37

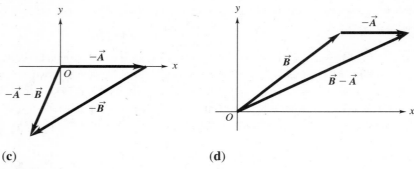

(c)   (d)

Figure 1.37

Reflect: $-\vec{A} - \vec{B} = -(\vec{A} + \vec{B})$ so it has the same magnitude and opposite direction as $\vec{A} + \vec{B}$. Similarly, $\vec{B} - \vec{A} = -(\vec{A} - \vec{B})$ and $\vec{B} - \vec{A}$ and $\vec{A} - \vec{B}$ have equal magnitudes and opposite directions.

1.41. Set Up: In each case, create a sketch (Figure 1.41) showing the components and the resultant to determine the quadrant in which the resultant vector $\vec{A}$ lies. The component vectors add to give the resultant.

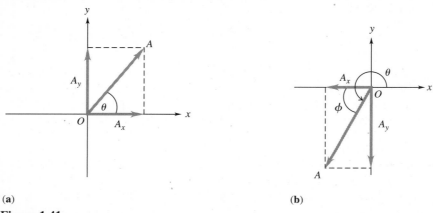

(a)   (b)

Figure 1.41

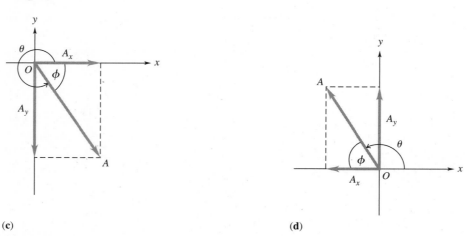

(c)   (d)

Figure 1.41

Solve: (a) $A = \sqrt{A_x^2 + A_y^2} = \sqrt{(4.0 \text{ m})^2 + (5.0 \text{ m})^2} = 6.4 \text{ m}$; $\tan\theta = \dfrac{A_y}{A_x} = \dfrac{5.0 \text{ m}}{4.0 \text{ m}}$ and $\theta = 51°$.

(b) $A = \sqrt{(-3.0 \text{ km})^2 + (-6.0 \text{ km})^2} = 6.7 \text{ km}$; $\tan\theta = \dfrac{-6.0 \text{ m}}{-3.0 \text{ m}}$ and $\theta = 243°$.

(My calculator gives $\phi = 63°$ and $\theta = \phi + 180°$.)

(c) $A = \sqrt{(9.0 \text{ m/s})^2 + (-17 \text{ m/s})^2} = 19 \text{ m/s}$; $\tan\theta = \dfrac{-17 \text{ m/s}}{9 \text{ m/s}}$ and $\theta = 298°$.

(My calculator gives $\phi = -62°$ and $\theta = \phi + 360°$.)

(d) $A = \sqrt{(-8.0 \text{ N})^2 + (12 \text{ N})^2} = 14 \text{ N}$. $\tan\theta = \dfrac{12 \text{ N}}{-8.0 \text{ N}}$ and $\theta = 124°$

(My calculator gives $\phi = -56°$ and $\theta = \phi + 180°$.)

Reflect: The signs of the components determine the quadrant in which the resultant lies.

1.45. Set Up: For parts (a) and (c), apply the appropriate signs to the relations $R_x = A_x + B_x$ and $R_y = A_y + B_y$. For (b) and (d), find the magnitude as $R = \sqrt{R_x^2 + R_y^2}$ and the direction as $\theta = \tan^{-1}(R_y/R_x)$.

Solve: (a) $R_x = A_x + B_x = 1.30 \text{ cm} + 4.10 \text{ cm} = 5.40 \text{ cm}$; $R_y = A_y + B_y = 2.25 \text{ cm} + (-3.75 \text{ cm}) = -1.50 \text{ cm}$

(b) $R = \sqrt{R_x^2 + R_y^2} = \sqrt{(5.40 \text{ cm})^2 + (-1.50 \text{ cm})^2} = 5.60 \text{ cm}$; $\theta = \tan^{-1}[(-1.50 \text{ cm})/(5.40 \text{ cm})] = -15.5°$. The resultant vector thus makes an angle of 344.5° counterclockwise from the $+x$ axis.

(c) $R_x = B_x + (-A_x) = 4.10 \text{ cm} + (-1.30 \text{ cm}) = 2.80 \text{ cm}$; $R_y = B_y + (-A_y) = -3.75 \text{ cm} + (-2.25 \text{ cm}) = -6.00 \text{ cm}$

(d) $R = \sqrt{(2.80 \text{ cm})^2 + (-6.00 \text{ cm})^2} = 6.62 \text{ cm}$; $\theta = \tan^{-1}[(-6.00 \text{ cm})/(2.80 \text{ cm})] = -65.0°$. The resultant vector thus makes an angle of 295.0° counterclockwise from the $+x$ axis.

Reflect: Note that $\vec{B} - \vec{A}$ has a larger magnitude than $\vec{B} + \vec{A}$. Vector addition is very different from addition of scalars.

1.47. Set Up: Use coordinates for which the $+y$ axis is downward. Let the two vectors be $\vec{A}$ and $\vec{B}$ and let the angles they make on either side of the $+y$ axis be ϕ_A and ϕ_B. Since the resultant, $\vec{R} = \vec{A} + \vec{B}$, is in the $+y$ direction, $R_x = 0$. The two vectors and their components are shown in Figure 1.47a.

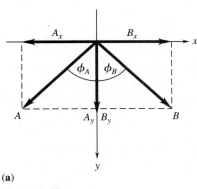 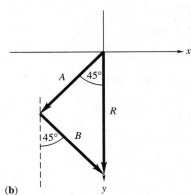

(a)

Figure 1.47

(b)

Solve: (a) $R_x = A_x + B_x = -A\sin\phi_A + B\sin\phi_B$. Since $A = B$ and $R_x = 0$, we can conclude that $\sin\phi_A = \sin\phi_B$ and $\phi_A = \phi_B$. Similarly, since $\phi_A + \phi_B = 90°$, $\phi_A = \phi_B = 45°$. The vector addition diagram for $\vec{R} = \vec{A} + \vec{B}$ is given in Figure 1.47b.

(b) $A_y + B_y = R_y$ and $R_y = 75 \text{ N}$. $A\cos\phi_A + B\cos\phi_B = 75 \text{ N}$. $A = B$ and $\phi_A = \phi_B = 45°$, so $2A(\cos 45°) = 75 \text{ N}$ and $A = 53 \text{ N}$.

Reflect: $A + B = 106 \text{ N}$. The magnitude of the vector sum $\vec{A} + \vec{B}$ is less than the sum of the magnitudes $A + B$.

1.49. Set Up: Use coordinates for which $+x$ is east and $+y$ is north. Each of the professor's displacement vectors make an angle of 0° or 180° with one of these axes. The components of his total displacement can thus be calculated directly from $R_x = A_x + B_x + C_x$ and $R_y = A_y + B_y + C_y$.

Solve: (a) $R_x = A_x + B_x + C_x = 0 + (-4.75 \text{ km}) + 0 = -4.75 \text{ km} = 4.75 \text{ km west}$; $R_y = A_y + B_y + C_y = 3.25 \text{ km} + 0 + (-1.50 \text{ km}) = 1.75 \text{ km} = 1.75 \text{ km north}$; $R = \sqrt{R_x^2 + R_y^2} = 5.06 \text{ km}$; $\theta = \tan^{-1}(R_y/R_x) = \tan^{-1}[(+1.75)/(-4.75)] = -20.2°$; $\phi = 180° - 20.2° = 69.8°$ west of north

(b) From the scaled sketch in Figure 1.49, the graphical sum agrees with the calculated values.

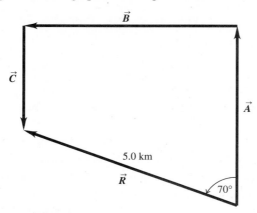

Figure 1.49

Reflect: The magnitude of his resultant displacement is very different from the distance he traveled, which is 159.50 km.

1.51. **Set Up:** We know that an object of mass 1000 g weighs 2.205 lbs and 16 oz = 1 lb.

Solve: $(5.00 \text{ oz})\left(\dfrac{1 \text{ lb}}{16 \text{ oz}}\right)\left(\dfrac{1000 \text{ g}}{2.205 \text{ lb}}\right) = 142 \text{ g}$ and $(5.25 \text{ oz})\left(\dfrac{1 \text{ lb}}{16 \text{ oz}}\right)\left(\dfrac{1000 \text{ g}}{2.205 \text{ lb}}\right) = 149 \text{ g}$

The acceptable limits are 142 g to 149 g.

Reflect: The range in acceptable weight is $\frac{1}{4}$ oz. Since $\frac{1}{4}$ oz = 7 g, the range in mass of 7 g is consistent. As we will see in a later chapter, mass has units of kg or g and is a different physical quantity than weight, which can have units of ounces. But for objects close to the surface of the earth, mass and weight are proportional; we can therefore say that a certain weight is equivalent to a certain mass.

1.53. **Set Up:** Estimate 12 breaths per minute. We know 1 day = 24 h, 1 h = 60 min and 1000 L = 1 m^3.

Solve: **(a)** $(12 \text{ breaths/min})\left(\dfrac{60 \text{ min}}{1 \text{ h}}\right)\left(\dfrac{24 \text{ h}}{1 \text{ day}}\right) = 17{,}280 \text{ breaths/day}$. The volume of air breathed in one day is $(\frac{1}{2} \text{ L/breath})(17{,}280 \text{ breaths/day}) = 8640 \text{ L} = 8.64 \text{ m}^3$.

The mass of air breathed in one day is the density of air times the volume of air breathed: $m = (1.29 \text{ kg/m}^3)(8.64 \text{ m}^3) = 11.1 \text{ kg}$. As 20% of this quantity is oxygen, the mass of oxygen breathed in 1 day is $(0.20)(11.1 \text{ kg}) = 2.2 \text{ kg} = 2200 \text{ g}$.

(b) 8.64 m^3 and $V = l^3$, so $l = V^{1/3} = 2.1$ m.

Reflect: A person could not survive one day in a closed tank of this size because the exhaled air is breathed back into the tank and thus reduces the percent of oxygen in the air in the tank. That is, a person cannot extract all of the oxygen from the air in an enclosed space.

1.59. **Set Up:** Calculate the volume of liquid in liters using 128 oz = 1 gal = 3.788 L. Use the result to calculate the mass of arsenic using the given arsenic per liter quantity 10 μg/L and the unit conversion 1 μg = 1 × 10^{-6} g.

Solve: You drink 64 oz = 2 qt = 0.5 gal or $(0.5 \text{ gal})[(3.788 \text{ L})/(1 \text{ gal})] = 1.894 \text{ L}$. The mass of arsenic received per day is $(10 \,\mu\text{g/L})(1.894 \text{ L}) = 19 \,\mu\text{g} = 1.9 \times 10^{-5} \text{ g}$.

Reflect: While this is a very small amount of arsenic, concern about the accumulation in the body over a long period of time may exist.

1.61. **Set Up:** Use coordinates for which $+x$ is east and $+y$ is north. The spelunker's vector displacements are: $\vec{A} = 180$ m, 0° of west; $\vec{B} = 210$ m, 45° east of south; $\vec{C} = 280$ m, 30° east of north; and the unknown displacement $\vec{D}$. The vector sum of these four displacements is zero.

Solve: $A_x + B_x + C_x + D_x = -180 \text{ m} + (210 \text{ m})(\sin 45°) + (280 \text{ m})(\sin 30°) + D_x = 0$ and $D_x = -108$ m. $A_y + B_y + C_y + D_y = (-210 \text{ m})(\cos 45°) + (280 \text{ m})(\cos 30°) + D_y = 0$ and $D_y = -94$ m.

$$D = \sqrt{D_x^2 + D_y^2} = 143 \text{ m}; \quad \theta = \tan^{-1}[(-94 \text{ m})/(-108)] = 41°; \quad \vec{D} = 143 \text{ m},\ 41° \text{ south of west.}$$

This result is confirmed by the sketch in Figure 1.61.

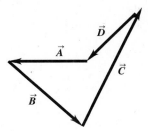

Figure 1.61

Reflect: We always add vectors by separately adding their x and y components.

1.65. Set Up: Referring to the vector diagram in Figure 1.65, the resultant weight of the forearm and the weight, $\vec{R}$, is 132.5 N upward while the biceps' force, $\vec{B}$, acts at 43° from the $+y$ axis. The force of the elbow is thus found as $\vec{E} = \vec{R} - \vec{B}$.

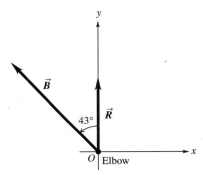

Figure 1.65

Solve: The components are: $E_x = R_x - B_x = 0 - (-232\,\text{N})(\sin 43°) = 158\,\text{N}$; $E_y = R_y - B_y = 132.5\,\text{N} - (232\,\text{N})(\cos 43°) = -37\,\text{N}$. The elbow force is thus

$$E = \sqrt{(158\,\text{N})^2 + (-37\,\text{N})^2} = 162\,\text{N}$$

and acts at $\theta = \tan^{-1}[(-37)/158] = -13°$ or 13° below the horizontal.

Reflect: The force exerted by the elbow is larger that the total weight of the arm and the object it is carrying.

2

MOTION ALONG A STRAIGHT LINE

Problems 1, 3, 5, 7, 17, 21, 23, 25, 29, 31, 33, 35, 39, 45, 47, 51, 53, 55, 57, 59, 63, 65, 69, 71, 75, 77, 81, 83, 87

Solutions to Problems

2.1. Set Up: Let the $+x$ direction be to the right in the figure.
Solve: **(a)** The lengths of the segments determine the distance of each point from O:

$$x_A = -5 \text{ cm}, \ x_B = +45 \text{ cm}, \ x_C = +15 \text{ cm}, \text{ and } x_D = -5 \text{ cm}.$$

(b) The displacement is Δx; the sign of Δx indicates its direction. The distance is always positive.
(i) A to B: $\Delta x = x_B - x_A = +45 \text{ cm} - (-5 \text{ cm}) = +50 \text{ cm}$. Distance is 50 cm.
(ii) B to C: $\Delta x = x_C - x_B = +15 \text{ cm} - 45 \text{ cm} = -30 \text{ cm}$. Distance is 30 cm.
(iii) C to D: $\Delta x = x_D - x_C = -5 \text{ cm} - 15 \text{ cm} = -20 \text{ cm}$. Distance is 20 cm.
(iv) A to D: $\Delta x = x_D - x_A = 0$. Distance $= 2(AB) = 100$ cm.
Reflect: When the motion is always in the same direction during the interval the magnitude of the displacement and the distance traveled are the same. In (iv) the ant travels to the right and then to the left and the magnitude of the displacement is less than the distance traveled.

2.3. Set Up: Let the $+x$ direction be to the right. $x_A = 2.0 \text{ m}, \ x_B = 7.0 \text{ m}, \ x_C = 6.0 \text{ m}.$
Solve: Average velocity is

$$v_{\text{av-}x} = \frac{\Delta x}{\Delta t} = \frac{x_C - x_A}{\Delta t} = \frac{+6.0 \text{ m} - 2.0 \text{ m}}{3.0 \text{ s}} = 1.3 \text{ m/s}.$$

$$\text{Average speed} = \frac{\text{distance}}{\text{time}} = \frac{4.0 \text{ m} + 1.0 \text{ m} + 1.0 \text{ m}}{3.0 \text{ s}} = 2.0 \text{ m/s}$$

Reflect: The average speed is greater than the magnitude of the average velocity.

2.5. Set Up: $x_A = 0, \ x_B = 3.0 \text{ m}, \ x_C = 9.0 \text{ m}. \ t_A = 0, \ t_B = 1.0 \text{ s}, \ t_C = 5.0 \text{ s}.$
Solve: **(a)** $v_{\text{av-}x} = \dfrac{\Delta x}{\Delta t}$

$$A \text{ to } B: v_{\text{av-}x} = \frac{\Delta x}{\Delta t} = \frac{x_B - x_A}{t_B - t_A} = \frac{3.0 \text{ m}}{1.0 \text{ s}} = 3.0 \text{ m/s}$$

$$B \text{ to } C: v_{\text{av-}x} = \frac{x_C - x_B}{t_C - t_B} = \frac{6.0 \text{ m}}{4.0 \text{ s}} = 1.5 \text{ m/s}$$

$$A \text{ to } C: v_{\text{av-}x} = \frac{x_C - x_A}{t_C - t_A} = \frac{9.0 \text{ m}}{5.0 \text{ s}} = 1.8 \text{ m/s}$$

(b) The velocity is always in the same direction ($+x$-direction), so the distance traveled is equal to the displacement in each case, and the average speed is the same as the magnitude of the average velocity.
Reflect: The average speed is different for different time intervals.

-1

2.7. Set Up: The positions x_t at time t are: $x_0 = 0$, $x_1 = 1.0$ m, $x_2 = 4.0$ m, $x_3 = 9.0$ m, $x_4 = 16.0$ m.

Solve: (a) The distance is $x_3 - x_1 = 8.0$ m.

(b) $v_{\text{av-}x} = \dfrac{\Delta x}{\Delta t}$. (i) $v_{\text{av},x} = \dfrac{x_1 - x_0}{1.0 \text{ s}} = 1.0$ m/s; (ii) $v_{\text{av},x} = \dfrac{x_2 - x_1}{1.0 \text{ s}} = 3.0$ m/s; (iii) $v_{\text{av},x} = \dfrac{x_3 - x_2}{1.0 \text{ s}} = 5.0$ m/s;

(iv) $v_{\text{av},x} = \dfrac{x_4 - x_3}{1.0 \text{ s}} = 7.0$ m/s; (v) $v_{\text{av},x} = \dfrac{x_4 - x_0}{4.0 \text{ s}} = 4.0$ m/s

Reflect: In successive 1 s time intervals the boulder travels greater distances and the average velocity for the intervals increases from one interval to the next.

2.17. Set Up: Use the normal driving time to find the distance. Use this distance to find the time on Friday.

Solve: $\Delta x = v_{\text{av},x}\Delta t = (105 \text{ km/h})(1.33 \text{ h}) = 140$ km. Then on Friday $\Delta t = \dfrac{\Delta x}{v_{\text{av},x}} = \dfrac{140 \text{ km}}{70 \text{ km/h}} = 2.00$ h. The

increase in time is $2.00 \text{ h} - 1.33 \text{ h} = 0.67 \text{ h} = 40$ min.

Reflect: A smaller average speed corresponds to a longer travel time when the distance is the same.

2.21. Set Up: Values of x_t at time t can be read from the graph: $x_0 = 0$, $x_4 = 3.0$ cm, $x_{10} = 4.0$ cm, and $x_{18} = 4.0$ cm. v_x is constant when x versus t is a straight line.

Solve: The motion consists of constant velocity segments.

$$t = 0 \text{ to } 4.0 \text{ s: } v_x = \frac{3.0 \text{ cm} - 0}{4.0 \text{ s}} = 0.75 \text{ cm/s};$$

$$t = 4.0 \text{ s to } 10.0 \text{ s: } v_x = \frac{4.0 \text{ cm} - 3.0 \text{ cm}}{6.0 \text{ s}} = 0.17 \text{ cm/s}; t = 10.0 \text{ s to } 18.0 \text{ s: } v_x = 0.$$

The graph of v_x versus t is shown in Figure 2.21.

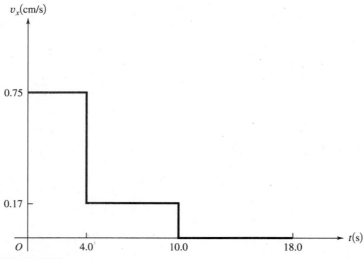

Figure 2.21

Reflect: v_x is the slope of x versus t.

2.23. Set Up: $a_{\text{av},x} = \dfrac{\Delta v_x}{\Delta t}$

Solve: (a) 0 s to 2 s: $a_{\text{av},x} = 0$; 2 s to 4 s: $a_{\text{av},x} = 1.0$ m/s²; 4 s to 6 s: $a_{\text{av},x} = 1.5$ m/s²; 6 s to 8 s: $a_{\text{av},x} = 2.5$ m/s²; 8 s to 10 s: $a_{\text{av},x} = 2.5$ m/s²; 10 s to 12 s: $a_{\text{av},x} = 2.5$ m/s²; 12 s to 14 s: $a_{\text{av},x} = 1.0$ m/s²; 14 s to 16 s: $a_{\text{av},x} = 0$. The acceleration is not constant over the entire 16 s time interval. The acceleration is constant between 6 s and 12 s.

(b) The graph of v_x versus t is given in Figure 2.23. $t = 9$ s: $a_x = 2.5$ m/s^2; $t = 13$ s: $a_x = 1.0$ m/s^2; $t = 15$ s: $a_x = 0$.

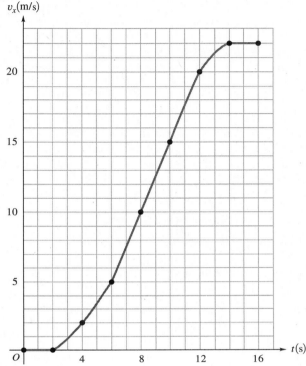

Figure 2.23

Reflect: The acceleration is constant when the velocity changes at a constant rate. When the velocity is constant, the acceleration is zero.

2.25. Set Up: The acceleration a_x equals the slope of the v_x versus t curve.
Solve: The qualitative graphs of acceleration as a function of time are given in Figure 2.25.

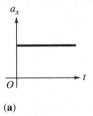 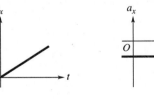 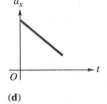

(a)   **(b)**   **(c)**   **(d)**

Figure 2.25

The acceleration can be described as follows: **(a)** positive and constant, **(b)** positive and increasing, **(c)** negative and constant, **(d)** positive and decreasing.

Reflect: When v_x and a_x have the same sign then the speed is increasing. In (c) the velocity and acceleration have opposite signs and the speed is decreasing.

2.29. Set Up: Assume constant acceleration. Take the $+x$ direction to be downward, in the direction of the motion of the capsule. $v_{0x} = 311$ km/h $= 86.4$ m/s, $v_x = 0$ (stops), $x_0 = 0$ and $x = 0.81$ m.

Solve: (a) $v_x^2 = v_{0x}^2 + 2a_x(x - x_0)$ and $a_x = \dfrac{v_x^2 - v_{0x}^2}{2(x - x_0)} = \dfrac{0 - (86.4 \text{ m/s})^2}{2(0.81 \text{ m})} = -4.61 \times 10^3 \text{ m/s}^2 = -470g$.

The minus sign tells us that a_x is upward.

(b) $v_x = v_{0x} + a_x t$ so $t = \dfrac{v_x - v_{0x}}{a_x} = \dfrac{0 - 86.4 \text{ m/s}}{-4.6 \times 10^3 \text{ m/s}^2} = 18.7$ ms

Reflect: Since the speed decreases, v_x and a_x must be in opposite directions.

2.31. Set Up: Assume your head accelerates upward 1.0 m in 0.25 s with constant acceleration. Let $+x$ be upward. $v_{0x} = 0$, $x - x_0 = 1.0$ m, and $t = 0.25$ s.

Solve: (a) $x - x_0 = v_{0x} t + \frac{1}{2} a_x t^2$ and $a_x = \dfrac{2(x - x_0)}{t^2} = \dfrac{2(1.0 \text{ m})}{(0.25 \text{ s})^2} = 30 \text{ m/s}^2 = 3g$

(b) No, different parts of your body travel through different distances in the same time.
Reflect: The shorter the time for you to stand up, the greater the acceleration.

2.33. Set Up: Let $+x$ be the direction the person travels. $v_x = 0$ (stops), $t = 36$ ms $= 3.6 \times 10^{-2}$ s, $a_x = -60g = -588 \text{ m/s}^2$. a_x is negative since it is opposite to the direction of the motion.
Solve: $v_x = v_{0x} + a_x t$ so $v_{0x} = -a_x t$. Then $x = x_0 + v_{0x} t + \frac{1}{2} a_x t^2$ gives $x = -\frac{1}{2} a_x t^2$.

$$x = -\tfrac{1}{2}(-588 \text{ m/s}^2)(3.6 \times 10^{-2} \text{ s})^2 = 38 \text{ cm}$$

Reflect: We could also find the initial speed: $v_{0x} = -a_x t = -(-588 \text{ m/s}^2)(36 \times 10^{-3} \text{ s}) = 21 \text{ m/s} = 47 \text{ mph}$

2.35. Set Up: Let $+x$ be in the direction of motion of the bullet. $v_{0x} = 0$, $x_0 = 0$, $v_x = 335 \text{ m/s}$, and $x = 0.127$ m.
Solve: (a) $v_x^2 = v_{0x}^2 + 2a_x(x - x_0)$ and

$$a_x = \frac{v_x^2 - v_{0x}^2}{2(x - x_0)} = \frac{(335 \text{ m/s})^2 - 0}{2(0.127 \text{ m})} = 4.42 \times 10^5 \text{ m/s}^2 = 4.51 \times 10^4 g$$

(b) $v_x = v_{0x} + a_x t$ so $t = \dfrac{v_x - v_{0x}}{a_x} = \dfrac{335 \text{ m/s} - 0}{4.42 \times 10^5 \text{ m/s}^2} = 0.758$ ms

Reflect: The acceleration is very large compared to g. In (b) we could also use $(x - x_0) = \left(\dfrac{v_{0x} + v_x}{2}\right) t$ to calculate
$t = \dfrac{2(x - x_0)}{v_x} = \dfrac{2(0.127 \text{ m})}{335 \text{ m/s}} = 0.758$ ms

2.39. Set Up: 0.250 mi $= 1320$ ft. 60.0 mph $= 88.0$ ft/s. Let $+x$ be the direction the car is traveling.
Solve: (a) braking: $v_{0x} = 88.0 \text{ ft/s}$, $x - x_0 = 146$ ft, $v_x = 0$. $v_x^2 = v_{0x}^2 + 2a_x(x - x_0)$ gives

$$a_x = \frac{v_x^2 - v_{0x}^2}{2(x - x_0)} = \frac{0 - (88.0 \text{ ft/s})^2}{2(146 \text{ ft})} = -26.5 \text{ ft/s}^2$$

Speeding up: $v_{0x} = 0$, $x - x_0 = 1320$ ft, $t = 19.9$ s. $x - x_0 = v_{0x} t + \frac{1}{2} a_x t^2$ gives

$$a_x = \frac{2(x - x_0)}{t^2} = \frac{2(1320 \text{ ft})}{(19.9 \text{ s})^2} = 6.67 \text{ ft/s}^2$$

(b) $v_x = v_{0x} + a_x t = 0 + (6.67 \text{ ft/s}^2)(19.9 \text{ s}) = 133 \text{ ft/s} = 90.5 \text{ mph}$
(c) $t = \dfrac{v_x - v_{0x}}{a_x} = \dfrac{0 - 88.0 \text{ ft/s}}{-26.5 \text{ ft/s}^2} = 3.32$ s
Reflect: The magnitude of the acceleration while braking is much larger than when speeding up. That is why it takes much longer to go from 0 to 60 mph than to go from 60 mph to 0.

2.45. Set Up: $a_A = a_B$, $x_{0A} = x_{0B} = 0$, $v_{0x,A} = v_{0x,B} = 0$, and $t_A = 2t_B$.

Solve: (a) $x = x_0 + v_{0x} t + \frac{1}{2} a_x t^2$ gives $x_A = \frac{1}{2} a_A t_A^2$ and $x_B = \frac{1}{2} a_B t_B^2$. $a_A = a_B$ gives $\dfrac{x_A}{t_A^2} = \dfrac{x_B}{t_B^2}$ and

$$x_B = \left(\frac{t_B}{t_A}\right)^2 x_A = (\tfrac{1}{2})^2 (250 \text{ km}) = 62.5 \text{ km}$$

(b) $v_x = v_{0x} + a_x t$ gives $a_A = \dfrac{v_A}{t_A}$ and $a_B = \dfrac{v_B}{t_B}$. Since $a_A = a_B$, $\dfrac{v_A}{t_A} = \dfrac{v_B}{t_B}$ and

$$v_B = \left(\frac{t_B}{t_A}\right) v_A = \left(\tfrac{1}{2}\right)(350\ \text{m/s}) = 175\ \text{m/s}.$$

Reflect: v_x is proportional to t and for $v_{0x} = 0$, x is proportional to t^2.

2.47. Set Up: $v_{0A} = v_{0B} = 0$. Let $x_{0A} = x_{0B} = 0$. $a_A = a_B$ and $v_B = 2v_A$.

Solve: (a) $v_x^2 = v_{0x}^2 + 2a_x(x - x_0)$ gives $a_x = \dfrac{v_x^2 - v_{0x}^2}{2(x - x_0)}$ and $\dfrac{v_A^2}{x_A} = \dfrac{v_B^2}{x_B}$.

$$x_B = \left(\frac{v_B}{v_A}\right)^2 x_A = 4(500\ \text{m}) = 2000\ \text{m}$$

(b) $v_x = v_{0x} + a_x t$ gives $a_x = \dfrac{v_x}{t} \cdot \dfrac{v_A}{t_A} = \dfrac{v_B}{t_B}$ and $t_B = \left(\dfrac{v_B}{v_A}\right) t_A = 2T$.

Reflect: x is proportional to v_x^2 and t is proportional to v_x.

2.51. Set Up: Take $+y$ upward. $v_y = 0$ at the maximum height. $a_y = -0.379g = -3.71\ \text{m/s}^2$.
Solve: Consider the motion from the maximum height back to the initial level. For this motion $v_{0y} = 0$ and
$t = 4.25\ \text{s}$. $y = y_0 + v_{0y}t + \tfrac{1}{2}a_y t^2 = \tfrac{1}{2}(-3.71\ \text{m/s}^2)(4.25\ \text{s})^2 = -33.5\ \text{m}$
The ball went 33.5 m above its original position.
(b) Consider the motion from just after it was hit to the maximum height. For this motion $v_y = 0$ and $t = 4.25\ \text{s}$.
$v_y = v_{0y} + a_y t$ gives $v_{0y} = -a_y t = -(-3.71\ \text{m/s}^2)(4.25\ \text{s}) = 15.8\ \text{m/s}$.
(c) The graphs are sketched in Figure 2.51.

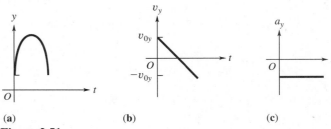

(a)   (b)   (c)

Figure 2.51

Reflect: The answers can be checked several ways. For example, $v_y = 0$, $v_{0y} = 15.8\ \text{m/s}$, and $a_y = -3.7\ \text{m/s}^2$ in
$v_y^2 = v_{0y}^2 + 2a_y(y - y_0)$ gives

$$y - y_0 = \frac{v_y^2 - v_{0y}^2}{2a_y} = \frac{0 - (15.8\ \text{m/s})^2}{2(-3.71\ \text{m/s}^2)} = 33.6\ \text{m},$$

which agrees with the height calculated in (a).

2.53. Set Up: Take $+y$ upward. $a_y = -9.80\ \text{m/s}^2$. The initial velocity of the sandbag equals the velocity of the balloon, so $v_{0y} = +5.00\ \text{m/s}$. When the balloon reaches the ground, $y - y_0 = -40.0\ \text{m}$. At its maximum height the sandbag has $v_y = 0$.
Solve: (a) $t = 0.250\ \text{s}$:

$$y - y_0 = v_{0y}t + \tfrac{1}{2}a_y t^2 = (5.00\ \text{m/s})(0.250\ \text{s}) + \tfrac{1}{2}(-9.80\ \text{m/s}^2)(0.250\ \text{s})^2 = 0.94\ \text{m}.$$

The sandbag is 40.9 m above the ground.

$$v_y = v_{0y} + a_y t = +5.00\ \text{m/s} + (-9.80\ \text{m/s}^2)(0.250\ \text{s}) = 2.55\ \text{m/s}.$$

$t = 1.00\ \text{s}$:

$$y - y_0 = (5.00\ \text{m/s})(1.00\ \text{s}) + \tfrac{1}{2}(-9.80\ \text{m/s}^2)(1.00\ \text{s})^2 = 0.10\ \text{m}.$$

The sandbag is 40.1 m above the ground. $v_y = v_{0y} + a_y t = +5.00 \text{ m/s} + (-9.80 \text{ m/s}^2)(1.00 \text{ s}) = -4.80 \text{ m/s}$.
(b) $y - y_0 = -40.0 \text{ m}$, $v_{0y} = 5.00 \text{ m/s}$, $a_y = -9.80 \text{ m/s}^2$. $y - y_0 = v_{0y}t + \frac{1}{2}a_y t^2$ gives $-40.0 \text{ m} = (5.00 \text{ m/s})t - (4.90 \text{ m/s}^2)t^2$. $(4.90 \text{ m/s}^2)t^2 - (5.00 \text{ m/s})t - 40.0 \text{ m} = 0$ and

$$t = \frac{1}{9.80}\left(5.00 \pm \sqrt{(-5.00)^2 - 4(4.90)(-40.0)}\right) \text{ s} = (0.51 \pm 2.90) \text{ s}.$$

t must be positive, so $t = 3.41 \text{ s}$.
(c) $v_y = v_{0y} + a_y t = +5.00 \text{ m/s} + (-9.80 \text{ m/s}^2)(3.41 \text{ s}) = -28.4 \text{ m/s}$
(d) $v_{0y} = 5.00 \text{ m/s}$, $a_y = -9.80 \text{ m/s}^2$, $v_y = 0$. $v_y^2 = v_{0y}^2 + 2a_y(y - y_0)$ gives

$$y - y_0 = \frac{v_y^2 - v_{0y}^2}{2a_y} = \frac{0 - (5.00 \text{ m/s})^2}{2(-9.80 \text{ m/s}^2)} = 1.28 \text{ m}.$$

The maximum height is 41.3 m above the ground.
(e) The graphs of a_y, v_y, and y versus t are given in Figure 2.53. Take $y = 0$ at the ground.

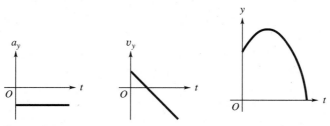

Figure 2.53

2.55. Set Up: $a_M = 0.170a_E$. Take $+y$ to be upward and $y_0 = 0$.
Solve: (a) $v_{0E} = v_{0M}$. $v_y^2 = v_{0y}^2 + 2a_y(y - y_0)$ with $v_y = 0$ at the maximum height gives $2a_y y = -v_{0y}^2$, so $a_M y_M = a_E y_E$.

$$y_M = \left(\frac{a_E}{a_M}\right)y_E = \left(\frac{1}{0.170}\right)(12.0 \text{ m}) = 70.6 \text{ m}$$

(b) Consider the time to the maximum height on the earth. The total travel time is twice this. First solve for v_{0y}, with $v_y = 0$ and $y = 12.0 \text{ m}$. $v_y^2 = v_{0y}^2 + 2a_y(y - y_0)$ gives

$$v_{0y} = \sqrt{-2(-a_E)y} = \sqrt{-2(-9.8 \text{ m/s}^2)(12.0 \text{ m})} = 15.3 \text{ m/s}.$$

Then $v_y = v_{0y} + a_y t$ gives

$$t = \frac{v_y - v_{0y}}{a_y} = \frac{0 - 15.3 \text{ m/s}}{-9.8 \text{ m/s}^2} = 1.56 \text{ s}.$$

The total time is $2(1.56 \text{ s}) = 3.12 \text{ s}$. Then, on the moon $v_y = v_{0y} + a_y t$ with $v_{0y} = 15.3 \text{ m/s}$, $v_y = 0$, and $a = -1.666 \text{ m/s}^2$ gives

$$t = \frac{v_y - v_{0y}}{a_y} = \frac{0 - 15.3 \text{ m/s}}{-1.666 \text{ m/s}^2} = 9.18 \text{ s}.$$

The total time is 18.4 s. It takes 15.3 s longer on the moon.
Reflect: The maximum height is proportional to $1/a$, so the height on the moon is greater. Since the acceleration is the rate of change of the speed, the wrench loses speed at a slower rate on the moon and it takes more time for its speed to reach $v = 0$ at the maximum height. In fact, $t_M/t_E = a_E/a_M = 1/0.170 = 5.9$, which agrees with our calculated times. But to find the difference in the times we had to solve for the actual times, not just their ratios.

2.57. Set Up: Take $+y$ upward. $a_y = -9.80 \text{ m/s}^2$. When the rock reaches the ground, $y - y_0 = -60.0 \text{ m}$.
Solve: (a) $y - y_0 = v_{0y}t + \frac{1}{2}a_yt^2$ gives $-60.0 \text{ m} = (12.0 \text{ m/s})t - (4.9 \text{ m/s}^2)t^2$. $(4.90 \text{ m/s}^2)t^2 - (12.0 \text{ m/s})t - 60.0 \text{ m} = 0$ and

$$t = \frac{1}{9.80}\left(12.0 \pm \sqrt{(-12.0)^2 - 4(4.90)(-60.0)}\right) \text{ s} = (1.22 \pm 3.71) \text{ s}.$$

t must be positive, so $t = 4.93 \text{ s}$.
(b) $v_y^2 = v_{0y}^2 + 2a_y(y - y_0)$ gives $v_y = -\sqrt{(12.0 \text{ m/s})^2 + 2(-9.80 \text{ m/s}^2)(-60.0 \text{ m})} = -36.3 \text{ m/s}$.
Reflect: We could have taken downward to be $+y$. Then $y - y_0$, v_y and a_y are all positive, but v_{0y} is negative. The same results are obtained with this alternative choice of coordinates.

2.59. Set Up: Use subscripts f and s to refer to the faster and slower stones, respectively. Take $+y$ to be upward and $y_0 = 0$ for both stones. $v_{0f} = 3v_{0s}$. When a stone reaches the ground, $y = 0$.
Solve: (a) $y = y_0 + v_{0y}t + \frac{1}{2}a_yt^2$ gives $a_y = -\dfrac{2v_{0y}}{t}$. Since both stones have the same a_y, $\dfrac{v_{0f}}{t_f} = \dfrac{v_{0s}}{t_s}$ and

$$t_s = t_f\left(\frac{v_{0s}}{v_{0f}}\right) = \left(\tfrac{1}{3}\right)10 \text{ s} = 3.3 \text{ s}$$

(b) Since $v_y = 0$ at the maximum height, then $v_y^2 = v_{0y}^2 + 2a_y(y - y_0)$ gives $a_y = -\dfrac{v_{0y}^2}{2y}$. Since both have the same a_y, $\dfrac{v_{0f}^2}{y_f} = \dfrac{v_{0s}^2}{y_s}$ and $y_f = y_s\left(\dfrac{v_{0f}}{v_{0s}}\right)^2 = 9H$.
Reflect: The faster stone reaches a greater height so it travels a greater distance than the slower stone and takes more time to return to the ground.

2.63. Set Up: Use coordinates with $+y$ downward. Relative to the earth the package has $v_{0y} = +3.50 \text{ m/s}$ and $a_y = 9.80 \text{ m/s}^2$.
Solve: The velocity of the package relative to the ground just before it hits is

$$v_y = \sqrt{v_{0y}^2 + 2a_y(y - y_0)} = \sqrt{(3.50 \text{ m/s})^2 + 2(9.80 \text{ m/s}^2)(8.50 \text{ m})} = 13.4 \text{ m/s}$$

(a) $\vec{v}_{P/G} = 13.4 \text{ m/s}$, downward. $\vec{v}_{H/G} = 3.50 \text{ m/s}$, downward

$$\vec{v}_{P/G} = \vec{v}_{P/H} + \vec{v}_{H/G} \text{ and } \vec{v}_{P/H} = \vec{v}_{P/G} - \vec{v}_{H/G}. \ \vec{v}_{P/H} = 9.9 \text{ m/s, downward.}$$

(b) $\vec{v}_{H/P} = -\vec{v}_{P/H}$, so $\vec{v}_{H/P} = 9.9 \text{ m/s}$, upward.
Reflect: Since the helicopter is traveling downward, the package is moving slower relative to the helicopter than its speed relative to the ground.

2.65. Set Up: The relative velocities are: $v_{P/G}$, the plane relative to the ground; $v_{P/A}$, the plane relative to the air; and $v_{A/G}$, the air relative to the ground. Let $+x$ be east.

Solve: Use the data for no wind to calculate $v_{P/A}$:

$$v_{P/A} = \frac{5310 \text{ km}}{6.60 \text{ h}} = 804.5 \text{ km/h}.$$

$v_{P/G,x} = v_{P/A,x} + v_{A/G,x}$. When flying east from A to B, $v_{P/A}$ and $v_{A/G}$ are both east and $v_{P/G} = 804.5 \text{ km/h} + v_{A/G}$.

$$t_{AB} = \frac{2655 \text{ km}}{v_{P/G}} = \frac{2655 \text{ km}}{804.5 \text{ km/h} + v_{A/G}}.$$

When flying west from B to A, $v_{P/A}$ is west and $v_{A/G}$ is east and $v_{P/G} = 804.5 \text{ km/h} - v_{A/G}$.

$$t_{BA} = \frac{2655 \text{ km}}{v_{P/G}} = \frac{2655 \text{ km}}{804.5 \text{ km/h} - v_{A/G}}.$$

$t_{AB} + t_{BA} = 6.70$ h, so

$$\frac{2655 \text{ km}}{804.5 \text{ km/h} + v_{A/G}} + \frac{2655 \text{ km}}{804.5 \text{ km/h} - v_{A/G}} = 6.70 \text{ h}$$

$$[(804.5 \text{ km/h})^2 - v_{A/G}^2][6.70 \text{ h}] = (2655 \text{ km})(1610 \text{ km/h}).$$

$$(804.5 \text{ km/h})^2 - v_{A/G}^2 = 6.376 \times 10^5 \text{ km}^2/\text{h}^2 \text{ and } v_{A/G} = 98.1 \text{ km/h}$$

Reflect: When the wind is blowing it increases the speed of the plane relative to the ground for the trip from A to B and decreases this speed for the return trip from B to A. Since the plane spends more time going from B to A than for A to B, the wind decreases the average speed for the roundtrip and therefore increases the total travel time.

2.69. Set Up: Let d be the distance from home to work. average speed $= \dfrac{d}{t}$

Solve: (a) The time to get to work is $t_1 = \dfrac{d}{60 \text{ mph}}$. The time to return home is $t_2 = \dfrac{d}{40 \text{ mph}}$. The average speed for the round trip is the total distance $2d$ divided by the total time $t_1 + t_2$:

$$\text{average speed} = \frac{2d}{t_1 + t_2} = 2\frac{(60)(40)}{60 + 40} \text{ mph} = 48 \text{ mph}$$

(b) The travel times at each speed are not the same. More time is spent traveling at 40 mph so the average speed is less than $(60 \text{ mph} + 40 \text{ mph})/2$.

Reflect: If the two average speeds for each one-way trip are v_1 and v_2, then the average speed for the roundtrip is $v_{av} = \dfrac{2v_1v_2}{v_1 + v_2}$. If $v_1 = v_2 = v$, then the average speed is this speed v. If the speeds differ a lot, then the average speed for the roundtrip differs greatly from $(v_1 + v_2)/2$. For example, if $v_1 = 90$ mph and $v_2 = 10$ mph, then $v_{av} = 18$ mph. In this case most of the time is spent while traveling at the slower speed.

2.71. Set Up: Let $+y$ to be upward and $y_0 = 0$. $a_M = a_E/6$. At the maximum height $v_y = 0$.
Solve: (a) $v_y^2 = v_{0y}^2 + 2a_y(y - y_0)$. Since $v_y = 0$ and v_{0y} is the same for both rocks, $a_M y_M = a_E y_E$ and

$$y_E = \left(\frac{a_M}{a_E}\right) y_M = H/6$$

(b) $v_y = v_{0y} + a_y t$. $a_M t_M = a_E t_E$ and $t_M = \left(\dfrac{a_E}{a_M}\right) t_E = 6(4.0 \text{ s}) = 24.0 \text{ s}$

Reflect: On the moon, where the acceleration is less, the rock reaches a greater height and takes more time to reach that maximum height.

2.75. Set Up: Let $+y$ be downward. The egg has $v_{0y} = 0$ and $a_y = 9.80 \text{ m/s}^2$. Find the distance the professor walks during the time t it takes the egg to fall to the height of his head. At this height, the egg has $y - y_0 = 44.2$ m.
Solve: $y - y_0 = v_{0y}t + \frac{1}{2}a_y t^2$ gives

$$t = \sqrt{\frac{2(y - y_0)}{a_y}} = \sqrt{\frac{2(44.2 \text{ m})}{9.80 \text{ m/s}^2}} = 3.00 \text{ s.}$$

The professor walks a distance $x - x_0 = v_{0x}t = (1.20 \text{ m/s})(3.00 \text{ s}) = 3.60$ m. Release the egg when your professor is 3.60 m from the point directly below you.
Reflect: Just before the egg lands its speed is $(9.80 \text{ m/s}^2)(3.00\text{s}) = 29.4 \text{ m/s}$. It is traveling much faster than the professor.

2.77. Set Up: Take $+y$ to be upward. **(a)** and **(b)** $v_{0y} = 4973 \text{ km/h} = 1381 \text{ m/s}$, $a_y = -9.8 \text{ m/s}^2$, and $y - y_0 = -45 \times 10^3$ m. **(c)** $v_{0y} = 0$, $v_y = 1381 \text{ m/s}$, and $y - y_0 = 45 \times 10^3$ m.

Solve: (a) $v_y^2 = v_{0y}^2 + 2a_y(y - y_0)$ and

$$v_y = -\sqrt{v_{0y}^2 + 2a_y(y - y_0)} = -\sqrt{(1381 \text{ m/s})^2 + 2(-9.8 \text{ m/s}^2)(-45 \times 10^3 \text{ m})}$$
$$v_y = -1670 \text{ m/s} = -6012 \text{ km/h}$$

(b) $v_y = v_{0y} + a_y t$ gives $t = \dfrac{v_y - v_{0y}}{a_y} = \dfrac{-1670 \text{ m/s} - 1381 \text{ m/s}}{-9.8 \text{ m/s}^2} = 310 \text{ s}$

(c) $v_y^2 = v_{0y}^2 + 2a_y(y - y_0)$ gives $a_y = \dfrac{v_y^2 - v_{0y}^2}{2(y - y_0)} = \dfrac{(1381 \text{ m/s})^2}{2(45 \times 10^3 \text{ m})} = 21 \text{ m/s}^2 = 2.2g$

$$y - y_0 = v_{0y}t + \tfrac{1}{2}a_y t^2 \text{ gives } t = \sqrt{\dfrac{2(y - y_0)}{a_y}} = \sqrt{\dfrac{2(45 \times 10^3 \text{ m})}{21 \text{ m/s}^2}} = 65 \text{ s}$$

Reflect: The SRBs have greater speed when they return to the ground than they had when released. After release they continue to travel upward. When they return to an altitude of 45 km on their way back down their speed is 4973 km/h, but they gain additional speed as they continue to fall to earth.

2.81. Set Up: Take $+y$ to be upward. There are two periods of constant acceleration: $a_y = +2.50 \text{ m/s}^2$ while the engines fire and $a_y = -9.8 \text{ m/s}^2$ after they shut off. Constant acceleration equations can be applied within each period of constant acceleration.

Solve: (a) Find the speed and height at the end of the first 20.0 s. $a_y = +2.50 \text{ m/s}^2$, $v_{0y} = 0$, and $y_0 = 0$. $v_y = v_{0y} + a_y t = (2.50 \text{ m/s}^2)(20.0 \text{ s}) = 50.0 \text{ m/s}$ and $y = y_0 + v_{0y}t + \tfrac{1}{2}a_y t^2 = \tfrac{1}{2}(2.50 \text{ m/s}^2)(20.0 \text{ s})^2 = 500 \text{ m}$. Next consider the motion from this point to the maximum height. $y_0 = 500 \text{ m}$, $v_y = 0$, $v_{0y} = 50.0 \text{ m/s}$, and $a_y = -9.8 \text{ m/s}^2$. $v_y^2 = v_{0y}^2 + 2a_y(y - y_0)$ gives

$$y - y_0 = \dfrac{v_y^2 - v_{0y}^2}{2a_y} = \dfrac{0 - (50.0 \text{ m/s})^2}{2(-9.8 \text{ m/s}^2)} = +128 \text{ m},$$

so $y = 628 \text{ m}$. The duration of this part of the motion is obtained from $v_y = v_{0y} + a_y t$:

$$t = \dfrac{v_y - v_{0y}}{a_y} = \dfrac{-50 \text{ m/s}}{-9.8 \text{ m/s}^2} = 5.10 \text{ s}$$

(b) At the highest point, $v_y = 0$ and $a_y = 9.8 \text{ m/s}^2$, downward.
(c) Consider the motion from the maximum height back to the ground. $a_y = -9.8 \text{ m/s}^2$, $v_{0y} = 0$, $y = 0$, and $y_0 = 628 \text{ m}$. $y = y_0 + v_{0y}t + \tfrac{1}{2}a_y t^2$ gives

$$t = \sqrt{\dfrac{2(y - y_0)}{a_y}} = 11.3 \text{ s}.$$

The total time the rocket is in the air is $20.0 \text{ s} + 5.10 \text{ s} + 11.3 \text{ s} = 36.4 \text{ s}$. $v_y = v_{0y} + a_y t = (-9.8 \text{ m/s}^2)(11.3 \text{ s}) = -111 \text{ m/s}$. Just before it hits the ground the rocket will have speed 111 m/s.
Reflect: We could calculate the time of free fall directly by considering the motion from the point of engine shutoff to the ground: $v_{0y} = 50.0 \text{ m/s}$, $y - y_0 = 500 \text{ m}$ and $a_y = -9.8 \text{ m/s}^2$. $y - y_0 = v_{0y}t + \tfrac{1}{2}a_y t^2$ gives $t = 16.4 \text{ s}$, which agrees with a total time of 36.4 s.

2.83. Set Up: The sign of v_x specifies the direction of motion. The slope of v_x versus t is the acceleration.
Solve: (a) The initial velocity is 20.0 cm/s, to the right.
(b) The mouse's greatest speed is 40.0 cm/s. When the mouse has this speed its velocity is positive so it is moving to the right.
(c) The mouse is moving to the right when v_x is positive. This is the case for $0 < t < 6.0 \text{ s}$ and for $9.0 \text{ s} < t < 10.0 \text{ s}$. The mouse is moving to the left when v_x is negative. This is the case for $6.0 \text{ s} < t < 9.0 \text{ s}$. The mouse is instantaneously at rest when $t = 6.0 \text{ s}$ and $t = 9.0 \text{ s}$.

(d) The mouse's largest magnitude of acceleration is between 5.0 s and about 6.3 s. The magnitude of the acceleration is

$$\frac{40.0 \text{ cm/s}}{1.0 \text{ s}} = 40.0 \text{ cm/s}^2.$$

The acceleration is negative so is directed to the left.

(e) The mouse changes direction when v_x changes sign. This happens at $t = 6.0$ s and 9.0 s.

(f) The mouse is speeding up when $|v_x|$ is increasing. This happens for $0 < t < 3.0$ s, $6.0 \text{ s} < t < 7.0 \text{ s}$ and for $9.0 \text{ s} < t < 10.0 \text{ s}$. The mouse is slowing down when $|v_x|$ is decreasing. This happens at $t = 3.0$ s, for $5.0 \text{ s} < t < 6.0 \text{ s}$ and for $7.0 \text{ s} < t < 9.0 \text{ s}$. The mouse is instantaneously at rest at $t = 6.0$ s and $t = 9.0$ s.

(g) No. These formulas apply only when the acceleration is constant. For constant acceleration the graph of v_x versus t is a straight line and that is not the case here.

Reflect: When v_x and a_x have the same sign, the speed is increasing. When v_x and a_x have opposite signs, the speed is decreasing. When $a_x = 0$ the speed is constant.

2.87. Set Up: Let t_{fall} be the time for the rock to fall to the ground and let t_s be the time it takes the sound to travel from the impact point back to you. $t_{\text{fall}} + t_s = 10.0$ s. Both the rock and sound travel a distance d that is equal to the height of the cliff. Take $+y$ downward for the motion of the rock. The rock has $v_{0y} = 0$ and $a_y = 9.80 \text{ m/s}^2$.

Solve: (a) For the rock, $y - y_0 = v_{0y}t + \frac{1}{2}a_yt^2$ gives $t_{\text{fall}} = \sqrt{\dfrac{2d}{9.80 \text{ m/s}^2}}$.

For the sound, $t_s = \dfrac{d}{330 \text{ m/s}} = 10.0$ s. Let $\alpha^2 = d$. $0.00303\alpha^2 + 0.4518\alpha - 10.0 = 0$. $\alpha = 19.6$ and $d = 384$ m.

(b) You would have calculated $d = \frac{1}{2}(9.80 \text{ m/s}^2)(10.0 \text{ s})^2 = 490$ m. You would have overestimated the height of the cliff. It actually takes the rock less time than 10.0 s to fall to the ground.

Reflect: Once we know d we can calculate that $t_{\text{fall}} = 8.8$ s and $t_s = 1.2$ s. The time for the sound of impact to travel back to you is 12% of the total time and cannot be neglected. The rock has speed 86 m/s just before it strikes the ground.

Motion in a Plane

Problems 3, 5, 9, 11, 13, 15, 19, 21, 23, 29, 35, 37, 41, 43, 49, 51, 53, 55, 57, 63, 65, 67, 69

Solutions to Problems

3.3. Set Up: Coordinates of point A are $(2.0\text{ m}, 1.0\text{ m})$ and for point B they are $(10.0\text{ m}, 6.0\text{ m})$.

Solve: (a) At A, $x = 2.0$ m, $y = 1.0$ m.

(b) $\vec{r}$ and its components are shown in Figure 3.3a. $r = \sqrt{x^2 + y^2} = 2.2$ m; $\tan\theta = \dfrac{y}{x}$ and $\theta = 26.6°$, counterclockwise from the $+x$-axis.

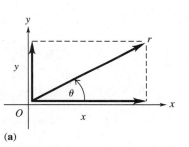

(a)

(b)

Figure 3.3

(c) $(v_x)_{\text{av}} = \dfrac{\Delta x}{\Delta t} = \dfrac{10.0\text{ m} - 2.0\text{ m}}{1.50\text{ s}} = 5.3\text{ m/s}$; $(v_y)_{\text{av}} = \dfrac{\Delta y}{\Delta t} = \dfrac{6.0\text{ m} - 1.0\text{ m}}{1.50\text{ s}} = 3.3\text{ m/s}$

(d) $\vec{v}_{\text{av}}$ and its components are shown in Figure 3.3b. $v_{\text{av}} = \sqrt{(v_x)_{\text{av}}^2 + (v_y)_{\text{av}}^2} = 6.2\text{ m/s}$; $\tan\phi = \dfrac{(v_y)_{\text{av}}}{(v_x)_{\text{av}}}$ and $\phi = 32°$, counterclockwise from the the the $+x$-axis.

Reflect: The displacement of the dragonfly is in the direction of $\vec{v}_{\text{av}}$.

3.5. Set Up: The coordinates of each point are: A, $(-50\text{ m}, 0)$; B, $(0, +50\text{ m})$; C, $(+50\text{ m}, 0)$; D, $(0, -50\text{ m})$. At each point the velocity is tangent to the circular path, as shown in Figure 3.5. The components (v_x, v_y) of the velocity at each point are: A, $(0, +6.0\text{ m/s})$; B, $(+6.0\text{ m/s}, 0)$; C, $(0, -6.0\text{ m/s})$; D, $(-6.0\text{ m/s}, 0)$.

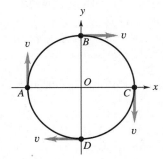

Figure 3.5

Solve: **(a)** *A to B* The time for one full lap is

$$t = \frac{2\pi r}{v} = \frac{2\pi(50 \text{ m})}{6.0 \text{ m/s}} = 52.4 \text{ s}.$$

A to B is one-quarter lap and takes $\frac{1}{4}(52.4 \text{ s}) = 13.1 \text{ s}$.

$$(v_x)_{av} = \frac{\Delta x}{\Delta t} = \frac{0 - (-50 \text{ m})}{13.1 \text{ s}} = 3.8 \text{ m/s}; (v_y)_{av} = \frac{\Delta y}{\Delta t} = \frac{+50 \text{ m} - 0}{13.1 \text{ s}} = 3.8 \text{ m/s}$$

$$(a_x)_{av} = \frac{\Delta v_x}{\Delta t} = \frac{6.0 \text{ m/s} - 0}{13.1 \text{ s}} = 0.46 \text{ m/s}^2; (a_y)_{av} = \frac{\Delta v_y}{\Delta t} = \frac{0 - 6.0 \text{ m/s}}{13.1 \text{ s}} = -0.46 \text{ m/s}^2$$

(b) *A to C* $t = \frac{1}{2}(52.4 \text{ s}) = 26.2 \text{ s}$

$$(v_x)_{av} = \frac{\Delta x}{\Delta t} = \frac{+50 \text{ m} - (-50 \text{ m})}{26.2 \text{ s}} = 3.8 \text{ m/s}; (v_y)_{av} = \frac{\Delta y}{\Delta t} = 0$$

$$(a_x)_{av} = \frac{\Delta v_x}{\Delta t} = 0; (a_y)_{av} = \frac{\Delta v_y}{\Delta t} = \frac{-6.0 \text{ m/s} - 6.0 \text{ m/s}}{26.2 \text{ s}} = -0.46 \text{ m/s}^2$$

(c) *C to D* $t = \frac{1}{4}(52.4 \text{ s}) = 13.1 \text{ s}$

$$(v_x)_{av} = \frac{\Delta x}{\Delta t} = \frac{0 - 50 \text{ m}}{13.1 \text{ s}} = -3.8 \text{ m/s}; (v_y)_{av} = \frac{\Delta y}{\Delta t} = \frac{-50 \text{ m} - 0}{13.1 \text{ s}} = -3.8 \text{ m/s}$$

$$(a_x)_{av} = \frac{\Delta v_x}{\Delta t} = \frac{-6.0 \text{ m/s} - 0}{13.1 \text{ s}} = -0.46 \text{ m/s}^2; (a_y)_{av} = \frac{\Delta v_y}{\Delta t} = \frac{0 - (-6.0 \text{ m/s})}{13.1 \text{ s}} = 0.46 \text{ m/s}^2$$

(d) *A to A* $\Delta x = \Delta y = 0$ so $(v_x)_{av} = (v_y)_{av} = 0$, and $\Delta v_x = \Delta v_y = 0$ so $(a_x)_{av} = (a_y)_{av} = 0$

(e) For *A to B*, $v_{av} = \sqrt{(v_x)_{av}^2 + (v_y)_{av}^2} = \sqrt{(3.8 \text{ m/s})^2 + (3.8 \text{ m/s})^2} \doteq 5.4 \text{ m/s}$. The speed is constant so the average speed is 6.0 m/s. The average speed is larger than the magnitude of the average velocity because the distance traveled is larger than the displacement.

(f) Velocity is a vector, with both magnitude and direction. The magnitude of the velocity is constant but its direction is changing.

Reflect: For this motion the acceleration describes the rate of change of the direction of the velocity.

3.9. Set Up: Take $+y$ downward, so $a_x = 0$, $a_y = +9.80 \text{ m/s}^2$ and $v_{0y} = 0$. When the ball reaches the floor, $y - y_0 = 0.750 \text{ m}$.

Solve: **(a)** $y - y_0 = v_{0y}t + \frac{1}{2}a_y t^2$ gives $t = \sqrt{\frac{2(y - y_0)}{a_y}} = \sqrt{\frac{2(0.750 \text{ m})}{9.80 \text{ m/s}^2}} = 0.391 \text{ s}$.

(b) $x - x_0 = v_{0x}t + \frac{1}{2}a_x t^2$ gives $v_{0x} = \frac{x - x_0}{t} = \frac{1.40 \text{ m}}{0.391 \text{ s}} = 3.58 \text{ m/s}$. Since $v_{0y} = 0$, $v_0 = v_{0x} = 3.58 \text{ m/s}$.

(c) $v_x = v_{0x} = 3.58 \text{ m/s}$. $v_y = v_{0y} + a_y t = (9.80 \text{ m/s}^2)(0.391 \text{ s}) = 3.83 \text{ m/s}$. $v = \sqrt{v_x^2 + v_y^2} = 5.24 \text{ m/s}$.

$$\tan\theta = \frac{|v_y|}{|v_x|} = \frac{3.83 \text{ m/s}}{3.58 \text{ m/s}}$$

and $\theta = 46.9°$. The final velocity of the ball has magnitude 5.24 m/s and is directed at 46.9° below the horizontal.

Reflect: The time for the ball to reach the floor is the same as if it had been dropped from a height of 0.750 m; the horizontal component of velocity has no effect on the vertical motion.

3.11. Set Up: Take $+y$ to be downward. For each cricket, $a_x = 0$ and $a_y = +9.80 \text{ m/s}^2$. For Milada, $v_{0x} = v_{0y} = 0$. For Chirpy, $v_{0x} = 0.950 \text{ m/s}$, $v_{0y} = 0$

Solve: Chirpy's horizontal component of velocity has no effect on his vertical motion. He also reaches the ground in 3.50 s. $x - x_0 = v_{0x}t + \frac{1}{2}a_x t^2 = (0.950 \text{ m/s})(3.50 \text{ s}) = 3.32 \text{ m}$

Reflect: The x and y components of motion are totally separate and are connected only by the fact that the time is the same for both.

3.13. Set Up: Take $+y$ to be downward. $a_x = 0$, $a_y = +9.80 \text{ m/s}^2$. $v_{0x} = v_0$, $v_{0y} = 0$. The car travels 21.3 m $-$ 1.80 m $= 19.5$ m downward during the time it travels 61.0 m horizontally.

Solve: Use the vertical motion to find the time in the air:

$$y - y_0 = v_{0y}t + \tfrac{1}{2}a_yt^2 \text{ gives } t = \sqrt{\frac{2(y - y_0)}{a_y}} = \sqrt{\frac{2(19.5 \text{ m})}{9.80 \text{ m/s}^2}} = 1.995 \text{ s}$$

Then $x - x_0 = v_{0x}t + \tfrac{1}{2}a_xt^2$ gives $v_0 = v_{0x} = \dfrac{x - x_0}{t} = \dfrac{61.0 \text{ m}}{1.995 \text{ s}} = 30.6 \text{ m/s}.$

(b) $v_x = 30.6 \text{ m/s}$ since $a_x = 0$. $v_y = v_{0y} + a_yt = -19.6 \text{ m/s}$. $v = \sqrt{v_x^2 + v_y^2} = 36.3 \text{ m/s}.$
Reflect: We calculate the final velocity by calculating its x and y components.

3.15. Set Up: Use coordinates with the origin at the initial position of the ball and $+y$ upward. $a_x = 0$ and $a_y = -9.80 \text{ m/s}^2$. The trajectory of the ball is sketched in Figure 3.15.

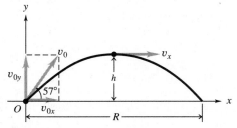

Figure 3.15

Solve: (a) From Figure 3.15, $v_{0x} = v_0\cos 57° = 13 \text{ m/s}$ and $v_{0y} = v_0\sin 57° = 20 \text{ m/s}.$
(b) At the maximum height $v_y = 0$, so $v_y^2 = v_{0y}^2 + 2a_y(y - y_0)$ with $y = h$ and $y_0 = 0$ gives

$$h = \frac{v_y^2 - v_{0y}^2}{2a_y} = \frac{0 - (20 \text{ m/s})^2}{2(-9.80 \text{ m/s}^2)} = 20 \text{ m}$$

(c) $v_y = v_{0y} + a_yt$ gives $t = \dfrac{v_y - v_{0y}}{a_y} = \dfrac{0 - 20 \text{ m/s}}{-9.80 \text{ m/s}^2} = 2.0 \text{ s}$

(d) $\vec{a} = g$, downward at all points of the trajectory. At the maximum height $v_y = 0$ and $v_x = v_{0x}$, so $\vec{v} = 13 \text{ m/s}$, horizontal.
(e) The total time in the air is twice the time to the maximum height, so is 4.0 s.
(f) $x - x_0 = v_{0x}t + \tfrac{1}{2}a_xt^2$ with $x = R$ and $x_0 = 0$ gives $R = v_{0x}t = (13 \text{ m/s})(4.0 \text{ s}) = 52 \text{ m}.$
Reflect: Our results for h and R agree with the general expressions derived in Example 3.5.

3.19. Set Up: The horizontal displacement when the ball returns to its original height is

$$R = \frac{v_0^2 \sin(2\theta_0)}{g}.$$

At its maximum height $v_y = 0$. $g = 32 \text{ ft/s}^2$. $a_x = 0$, $a_y = -g$.

Solve: (a) $v_0 = \sqrt{\dfrac{Rg}{\sin(2\theta_0)}} = \sqrt{\dfrac{(375 \text{ ft})(32 \text{ ft/s}^2)}{\sin(60.0°)}} = 118 \text{ ft/s}$

(b) $v_{0y} = v_0\sin\theta_0 = (118 \text{ ft/s})\sin 30.0° = 59 \text{ ft/s}$. $v_y^2 = v_{0y}^2 + 2a_y(y - y_0)$.

$$y - y_0 = \frac{v_y^2 - v_{0y}^2}{2a_y} = \frac{0 - (59 \text{ ft/s})^2}{2(-32 \text{ ft/s}^2)} = 54.4 \text{ ft}$$

Reflect: At the maximum height $v_y = 0$. But $v \neq 0$ there because the ball still has its constant horizontal component of velocity. The horizontal range equation,

$$R = \frac{v_0^2 \sin(2\theta_0)}{g},$$

can be used only when the initial and final points of the motion are at the same elevation.

3.21. Set Up: Use coordinates with the origin at the ground and $+y$ upward. $a_x = 0$, $a_y = -9.80 \text{ m/s}^2$. At the maximum height $v_y = 0$.
Solve: (a) $v_y^2 = v_{0y}^2 + 2a_y(y - y_0)$ gives

$$v_y = \sqrt{-2a_y(y - y_0)} = \sqrt{-2(-9.80 \text{ m/s}^2)(0.587 \text{ m})} = 3.39 \text{ m/s}$$

$$v_{0y} = v_0 \sin\theta_0 \text{ so } v_0 = \frac{v_{0y}}{\sin\theta_0} = \frac{3.39 \text{ m/s}}{\sin 58.0°} = 4.00 \text{ m/s}$$

(b) Use the vertical motion to find the time in the air. When the froghopper has returned to the ground, $y - y_0 = 0$.

$$y - y_0 = v_{0y}t + \tfrac{1}{2}a_y t^2 \text{ gives } t = -\frac{2v_{0y}}{a_y} = -\frac{2(3.39 \text{ m/s})}{-9.80 \text{ m/s}^2} = 0.692 \text{ s}$$

Then $x - x_0 = v_{0x}t + \tfrac{1}{2}a_x t^2 = (v_0 \cos\theta_0)t = (4.00 \text{ m/s})(\cos 58.0°)(0.692 \text{ s}) = 1.47 \text{ m}$

Reflect: $v_y = 0$ when $t = -\dfrac{v_{0y}}{a_y} = -\dfrac{3.39 \text{ m/s}}{-9.80 \text{ m/s}^2} = 0.346$ s. The total time in the air is twice this.

3.23. Set Up: Let $+y$ be upward. $a_x = 0$, $a_y = -9.80 \text{ m/s}^2$. $v_{0x} = v_0 \cos\theta_0$, $v_{0y} = v_0 \sin\theta_0$.
Solve: (a) $x - x_0 = v_{0x}t + \tfrac{1}{2}a_x t^2$ gives $x - x_0 = v_0(\cos\theta_0)t$ and

$$\cos\theta_0 = \frac{45.0 \text{ m}}{(25.0 \text{ m/s})(3.00 \text{ s})} = 0.600; \ \theta_0 = 53.1°$$

(b) At the highest point $v_x = v_{0x} = (25.0 \text{ m/s})\cos 53.1° = 15.0 \text{ m/s}$, $v_y = 0$ and $v = \sqrt{v_x^2 + v_y^2} = 15.0 \text{ m/s}$. At all points in the motion, $a = 9.80 \text{ m/s}^2$ downward.
(c) Find $y - y_0$ when $t = 3.00$ s:

$$y - y_0 = v_{0y}t + \tfrac{1}{2}a_y t^2 = (25.0 \text{ m/s})(\sin 53.1°)(3.00 \text{ s}) + \tfrac{1}{2}(-9.80 \text{ m/s}^2)(3.00 \text{ s})^2 = 15.9 \text{ m}$$

$$v_x = v_{0x} = 15.0 \text{ m/s}, v_y = v_{0y} + a_y t = (25.0 \text{ m/s})(\sin 53.1°) - (9.80 \text{ m/s}^2)(3.00 \text{ s}) = -9.41 \text{ m/s, and}$$

$$v = \sqrt{v_x^2 + v_y^2} = \sqrt{(15.0 \text{ m/s})^2 + (-9.41 \text{ m/s})^2} = 17.7 \text{ m/s}$$

Reflect: The acceleration is the same at all points of the motion. It takes the water

$$t = -\frac{v_{0y}}{a_y} = -\frac{20.0 \text{ m/s}}{-9.80 \text{ m/s}^2} = 2.04 \text{ s}$$

to reach its maximum height. When the water reaches the building it has passed its maximum height and its vertical component of velocity is downward.

3.29. Set Up: Example 3.5 derives $R = \dfrac{v_0^2 \sin 2\theta_0}{g}$.

Solve: Use the 66.0° data:

$$\frac{v_0^2}{g} = \frac{R}{\sin 2\theta_0} = \frac{275 \text{ m}}{\sin 132.0°} = 370 \text{ m.}$$

R_{max} is for $\theta_0 = 45.0°$ and $R_{max} = \dfrac{v_0^2}{g} = 370$ m.

Reflect: R is maximum when $\theta_0 = 45.0°$ and decreases as the angle moves away from 45.0°. $R \rightarrow 0$ as $\theta_0 \rightarrow 0$ or $\theta_0 \rightarrow 90.0°$.

3.35. Set Up: $R = 0.150$ m. $T = 1$ min $= 60.0$ s.

Solve: $a_{\text{rad}} = \dfrac{4\pi^2 R}{T^2} = \dfrac{4\pi^2(0.150 \text{ m})}{(60.0 \text{ s})^2} = 1.64 \times 10^{-3}$ m/s^2

Reflect: Different points in the hand travel in circles of different radii but have the same period so they have different a_{rad}. a_{rad} is largest for the tip, since it has the largest R.

3.37. Set Up: $R = 0.070$ m. For 3.0 rev/s, the period T (time for one revolution) is $T = (1.0 \text{ s})/(3.0 \text{ rev}) = 0.333$ s.

Solve: $a_{\text{rad}} = \dfrac{4\pi^2 R}{T^2} = \dfrac{4\pi^2(0.070 \text{ m})}{(0.333 \text{ s})^2} = 25$ m/s$^2 = 2.5g$

Reflect: The acceleration is large and the force on the fluid must be 2.5 times its weight.

3.41. Set Up: The relative velocities are the water relative to the earth, $\vec{v}_{\text{W/E}}$, the boat relative to the water, $\vec{v}_{\text{B/W}}$, and the boat relative to the earth, $\vec{v}_{\text{B/E}}$. $\vec{v}_{\text{B/E}}$ is due east, $\vec{v}_{\text{W/E}}$ is due south and has magnitude 2.0 m/s. $v_{\text{B/W}} = 4.2$ m/s. $\vec{v}_{\text{B/E}} = \vec{v}_{\text{B/W}} + \vec{v}_{\text{W/E}}$. The velocity addition diagram is given in Figure 3.41.

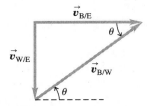

Figure 3.41

Solve: (a) Find the direction of $\vec{v}_{\text{B/W}}$. $\sin\theta = \dfrac{v_{\text{W/E}}}{v_{\text{B/W}}} = \dfrac{2.0 \text{ m/s}}{4.2 \text{ m/s}}$. $\theta = 28.4°$, north of east.

(b) $v_{\text{B/E}} = \sqrt{v_{\text{B/W}}^2 - v_{\text{W/E}}^2} = \sqrt{(4.2 \text{ m/s})^2 - (2.0 \text{ m/s})^2} = 3.7$ m/s

(c) $t = \dfrac{800 \text{ m}}{v_{\text{B/E}}} = \dfrac{800 \text{ m}}{3.7 \text{ m/s}} = 216$ s.

Reflect: It takes longer to cross the river in this problem than it did in Problem 3.40. In the direction straight across the river (east) the component of his velocity relative to the earth is lass than 4.2 m/s.

3.43. Set Up: $v_{\text{B/A}} = 100$ km/h. $\vec{v}_{\text{A/G}} = 40$ km/h, east.

Solve: $\vec{v}_{\text{B/G}} = \vec{v}_{\text{B/A}} + \vec{v}_{\text{A/G}}$. We want $\vec{v}_{\text{B/G}}$ to be due south. The relative velocity addition diagram is shown in Figure 3.43.

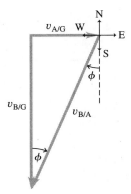

Figure 3.43

(a) $\sin\phi = \dfrac{v_{\text{A/G}}}{v_{\text{B/A}}} = \dfrac{40 \text{ km/h}}{100 \text{ km/h}}$, $\phi = 24°$, west of south.

(b) $v_{B/G} = \sqrt{v_{B/A}^2 - v_{A/G}^2} = 91.7 \text{ km/h}.$ $t = \dfrac{d}{v_{B/G}} = \dfrac{500 \text{ km}}{91.7 \text{ km/h}} = 5.5 \text{ h}$

Reflect: The speed of the bird relative to the ground is less than its speed relative to the air. Part of its velocity relative to the air is directed to oppose the effect of the wind.

3.49. Set Up: Use coordinates with $+y$ upward and $+x$ horizontal. The bale has $v_{0x} = v_0\cos\theta_0 = (75 \text{ m/s})\cos 55° = 43.0 \text{ m/s}$ and $v_{0y} = v_0\sin\theta_0 = (75 \text{ m/s})\sin 55° = 61.4 \text{ m/s}.$ $y_0 = 150 \text{ m}$ and $y = 0$.

Solve: Use the vertical motion to find t: $y - y_0 = v_{0y}t + \frac{1}{2}a_y t^2$ gives $-150 \text{ m} = (61.4 \text{ m/s})t - (4.90 \text{ m/s}^2)t^2$. The quadratic formula gives $t = 6.27 \pm 8.36 \text{ s}$. The physical value is the positive one, and $t = 14.6 \text{ s}$. Then $x - x_0 = v_{0x}t + \frac{1}{2}a_x t^2 = (43.0 \text{ m/s})(14.6 \text{ s}) = 630 \text{ m}$.

Reflect: If the airplane maintains constant velocity after it releases the bales, it will also travel horizontally 630 m during the time it takes the bales to fall to the ground, so the airplane will be directly over the impact spot when the bales land.

3.51. Set Up: Take $+y$ upward. For the balloon, $v_{0x} = v_0\cos\theta_0 = (12.0 \text{ m/s})\cos 50.0° = 7.71 \text{ m/s}$ and $v_{0y} = v_0\sin\theta_0 = (12.0 \text{ m/s})\sin 50.0° = 9.19 \text{ m/s}$.

Solve: Use the vertical motion of the balloon to find the time it is in the air: $y - y_0 = v_{0y}t + \frac{1}{2}a_y t^2$ with $y - y_0 = 0$ gives

$$t = -\frac{2v_{0y}}{a_y} = -\frac{2(9.19 \text{ m/s})}{-9.80 \text{ m/s}^2} = 1.88 \text{ s}.$$

The balloon travels a horizontal distance $x - x_0 = v_{0x}t + \frac{1}{2}a_x t^2 = (7.71 \text{ m/s})(1.88 \text{ s}) = 14.5 \text{ m}$. During this time the car travels a distance $d = (8.00 \text{ m/s})(1.88 \text{ s}) = 15.0 \text{ m}$. The car must be within $14.5 \text{ m} + 15.0 \text{ m} = 29.5 \text{ m}$ from the girl when she throws the balloon.

Reflect: The angle at which she throws the balloon affects both its horizontal range and its time in the air.

3.53. Set Up: Example 3.5 gives $R = \dfrac{v_0^2 \sin 2\theta_0}{g}$, $h = \dfrac{v_0^2 \sin\theta_0}{2g}$ and $t = \dfrac{2v_0 \sin\theta_0}{g}$.

Solve: (a) $R_{max} = \dfrac{v_0^2}{g}$ and occurs for $\theta_0 = 45°$.

$$g = \frac{v_0^2}{R_{max}} = \frac{(17.6 \text{ m/s})^2}{33.8 \text{ m}} = 9.16 \text{ m/s}^2$$

(b) $t = \dfrac{2v_0 \sin\theta_0}{g} = \dfrac{2(17.6 \text{ m/s})\sin 45°}{9.16 \text{ m/s}^2} = 2.72 \text{ s}$

(c) $h = \dfrac{v_0^2 \sin\theta_0}{2g} = \dfrac{(17.6 \text{ m/s})^2 (\sin 45°)^2}{2(9.16 \text{ m/s}^2)} = 8.45 \text{ m}$

Reflect: All three of these quantities, the range, the time in the air and the maximum height, are less than they would be on earth with the same initial speed, because g is larger on this planet than it is on earth.

3.55. Set Up: Let $+y$ be upward. $a_x = 0$, $a_y = -9.80 \text{ m/s}^2$. With $x_0 = y_0 = 0$, x and y are related by

$$y = (\tan\theta_0)x - \frac{g}{2v_0^2 \cos^2\theta_0}x^2.$$

$\theta_0 = 60.0°$. $y = 8.00 \text{ m}$ when $x = 18.0 \text{ m}$.

Solve: (a) $v_0 = \sqrt{\dfrac{gx^2}{2(\cos^2\theta_0)(x\tan\theta_0 - y)}} = 16.6 \text{ m/s}$

(b) $v_x = v_{0x} = v_0\cos\theta_0 = 8.3$ m/s. $v_y^2 = v_{0y}^2 + 2a_y(y - y_0)$ gives

$$v_y = -\sqrt{(v_0\sin\theta_0)^2 + 2a_y(y - y_0)} = -\sqrt{(14.4 \text{ m/s})^2 + 2(-9.80 \text{ m/s}^2)(8.00 \text{ m})} = -7.1 \text{ m/s}$$

$v = \sqrt{v_x^2 + v_y^2} = 10.9$ m/s. $\tan\theta = \dfrac{|v_y|}{|v_x|} = \dfrac{7.1}{8.3}$ and $\theta = 40.5°$, below the horizontal.

Reflect: We can check our calculated v_0.

$$t = \frac{x - x_0}{v_{0x}} = \frac{18.0 \text{ m}}{8.3 \text{ m/s}} = 2.17 \text{ s}.$$

Then $y - y_0 = v_{0y}t + \frac{1}{2}a_yt^2 = (14.4 \text{ m/s})(2.17 \text{ s}) - (4.9 \text{ m/s}^2)(2.17 \text{ s})^2 = 8$ m, which checks.

3.57. Set Up: Use coordinates with the origin at the boy and with $+y$ upward. The ball has $v_{0x} = v_0\cos\theta_0 = (8.50 \text{ m/s})\cos 60.0° = 4.25$ m/s, $v_{0y} = v_0\sin\theta_0 = (8.50 \text{ m/s})\sin 60.0° = 7.36$ m/s, $a_x = 0$ and $a_y = -9.80$ m/s^2.
Solve: (a) The dog must travel horizontally the same distance the ball travels horizontally, so the dog must have speed 4.25 m/s.
(b) Use the vertical motion of the ball to find its time in the air. $y - y_0 = v_{0y}t + \frac{1}{2}a_yt^2$ gives

$$-12.0 \text{ m} = (7.36 \text{ m/s})t - (4.90 \text{ m/s}^2)t^2.$$

The quadratic formula gives $t = 0.751 \pm 1.74$ s. The negative value is unphysical, so $t = 2.49$ s. Then $x - x_0 = v_{0x}t + \frac{1}{2}a_xt^2 = (4.25 \text{ m/s})(2.49 \text{ s}) = 10.6$ m
Reflect: The ball is in the air longer than when it is thrown horizontally (Problem 3.56), but it doesn't travel as far horizontally. The dog doesn't have to run as far or as fast as in Problem 3.56.

3.63. Set Up: Use coordinates with the origin at the ground and $+y$ upward. The shot put has $y_0 = 2.00$ m, $v_{0x} = v_0\cos\theta_0$, $v_{0y} = v_0\sin\theta_0$, $a_x = 0$ and $a_y = -9.80$ m/s^2. 1 mph = 0.4470 m/s
Solve: $x - x_0 = v_{0x}t + \frac{1}{2}a_xt^2$ gives 23.11 m $= (v_0\cos 40.0°)t$ and $v_0t = 30.17$ m

$$y - y_0 = v_{0y}t + \frac{1}{2}a_yt^2 \text{ gives } 0 = 2.00 \text{ m} + (v_0\sin 40.0°)t - (4.90 \text{ m/s}^2)t^2.$$

Use $v_0t = 30.17$ m and solve for t. This gives $t = 2.09$ s. Then $v_0 = \dfrac{30.17 \text{ m}}{2.09 \text{ s}} = 14.4$ m/s $= 32.2$ mph

Reflect: Since the initial and final heights are not the same, the expression $R = \dfrac{v_0^2\sin(2\theta_0)}{g}$ does not apply.

3.65. Set Up: Let $+y$ be upward. $a_x = 0$, $a_y = -9.80$ m/s^2. With $x_0 = y_0 = 0$, x and y are related by

$$y = (\tan\theta_0)x - \frac{g}{2v_0^2\cos^2\theta_0}x^2.$$

Solve: (a) $x = 40.0$ m when $y = -15.0$ m.

$$v_0 = \sqrt{\frac{gx^2}{2(\cos^2\theta_0)(x\tan\theta_0 - y)}} = \sqrt{\frac{(9.80 \text{ m/s}^2)(40.0 \text{ m})^2}{2(\cos 53.0°)^2([40.0 \text{ m}]\tan 53.0° + 15.0 \text{ m})}} = 17.8 \text{ m/s}$$

(b) Find x for $v_0 = 8.9$ m/s and $y = -100$ m.

$$y = (\tan\theta_0)x - \frac{g}{2v_0^2\cos^2\theta_0}x^2 \text{ gives } -100 = 1.33x - 0.171x^2.$$

The quadratic formula gives $x = 28.4$ m. He lands in the river 28.4 m from the near bank.
Reflect: Before doing the calculation for part (b) we don't know that he landed in the river. If we had found $x > 40$ m for $y = -100$ m, we would know he hit the vertical face of the far bank before he reached the water.

3.67. Set Up: Take $+y$ downward. $v_{0x} = v_0$, $v_{0y} = 0$. $a_x = 0$, $a_y = +9.80 \text{ m/s}^2$.

Solve: **(a)** Use the vertical motion to find the time for the boulder to reach the level of the lake: $y - y_0 = v_{0y}t + \frac{1}{2}a_yt^2$ with $y - y_0 = +20 \text{ m}$ gives

$$t = \sqrt{\frac{2(y - y_0)}{a_y}} = \sqrt{\frac{2(20 \text{ m})}{9.80 \text{ m/s}^2}} = 2.02 \text{ s}.$$

The rock must travel horizontally 100 m during this time. $x - x_0 = v_{0x}t + \frac{1}{2}a_xt^2$ gives

$$v_0 = v_{0x} = \frac{x - x_0}{t} = \frac{100 \text{ m}}{2.02 \text{ s}} = 49 \text{ m/s}$$

(b) In going from the edge of the cliff to the plain, the boulder travels downward a distance of $y - y_0 = 45 \text{ m}$.

$$t = \sqrt{\frac{2(y - y_0)}{a_y}} = \sqrt{\frac{2(45 \text{ m})}{9.80 \text{ m/s}^2}} = 3.03 \text{ s}$$

and $x - x_0 = v_{0x}t = (49 \text{ m/s})(3.03 \text{ s}) = 148 \text{ m}$. The rock lands $148 \text{ m} - 100 \text{ m} = 48 \text{ m}$ beyond the foot of the dam.

Reflect: The boulder passes over the dam 2.02 s after it leaves the cliff and then travels an additional 1.01 s before landing on the plain. If the boulder has an initial speed that is less than 49 m/s, then it lands in the lake.

3.69. Set Up: Use coordinates with $+y$ upward. Both the shell and the projectile have $a_x = 0$ and $a_y = -9.80 \text{ m/s}^2$ as they move through the air.

Solve: **(a)** Find the maximum height of the shell: $v_{0y} = v_0\sin\theta_0 = 120 \text{ m/s}$ and $v_y = 0$

$$v_y^2 = v_{0y}^2 + 2a_y(y - y_0) \text{ gives } y - y_0 = \frac{v_y^2 - v_{0y}^2}{2a_y} = \frac{0 - (120 \text{ m/s})^2}{2(-9.80 \text{ m/s}^2)} = 735 \text{ m}$$

At its maximum height the shell has $v_x = v_{0x} = v_0\cos 53° = 90.3 \text{ m/s}$ and $v_y = 0$. Relative to the ground the projectile has velocity components

$$v_{0x} = 90.3 \text{ m/s} + (100.0 \text{ m/s})\cos 30.0° = 176.9 \text{ m/s and}$$

$$v_{0y} = (100.0 \text{ m/s})\sin 30.0° = 50.0 \text{ m/s}$$

Find the maximum height of the projectile: $y_0 = 735 \text{ m}$. $v_y = 0$. $v_y^2 = v_{0y}^2 + 2a_y(y - y_0)$ gives

$$y - y_0 = \frac{v_y^2 - v_{0y}^2}{2a_y} = \frac{0 - (50.0 \text{ m/s})^2}{2(-9.80 \text{ m/s}^2)} = 128 \text{ m}$$

and the maximum height is $y = 735 \text{ m} + 128 \text{ m} = 863 \text{ m}$.

(b) Find the time for the shell to reach its maximum height: $v_y = v_{0y} + a_yt$ gives

$$t = \frac{v_y - v_{0y}}{a_y} = \frac{0 - 120 \text{ m/s}}{-9.80 \text{ m/s}^2} = 12.2 \text{ s}$$

During this time the shell has traveled horizontally a distance

$$x - x_0 = v_{0x}t = (v_0\cos\theta_0)t = (150 \text{ m/s})(\cos 53°)(12.2 \text{ s}) = 1100 \text{ m}$$

Find the time for the projectile to reach the ground from the point where it is launched. It starts at $y_0 = 735 \text{ m}$ and ends up at $y = 0$, so $y - y_0 = -735 \text{ m}$. $y - y_0 = v_{0y}t + \frac{1}{2}a_yt^2$ gives $-735 \text{ m} = (50.0 \text{ m/s})t - (4.90 \text{ m/s}^2)t^2$ and $t = 18.4 \text{ s}$. During this time the projectile has a horizontal displacement $x - x_0 = v_{0x}t = (176.9 \text{ m/s})(18.4 \text{ s}) = 3255 \text{ m}$. It is at $x = x_0 + 3255 \text{ m} = 1100 \text{ m} + 3255 \text{ m} = 4360 \text{ m}$ from the original launch point when it reaches the ground.

Reflect: If the shell didn't fire the projectile its horizontal range would be

$$R = \frac{v_0^2\sin(2\theta_0)}{g} = \frac{(150 \text{ m/s})^2(\sin 106°)}{9.80 \text{ m/s}^2} = 2210 \text{ m}.$$

The projectile gets an extra boost when it is fired and travels farther than this.

Newton's Laws of Motion

Problems 3, 5, 9, 11, 17, 19, 21, 23, 27, 29, 35, 37, 39, 43, 45, 47, 49, 51, 57

Solutions to Problems

4.3. Set Up: The force and its horizontal and vertical components are shown in Figure 4.3. The force is directed at an angle of $20.0° + 30.0° = 50.0°$ above the horizontal.

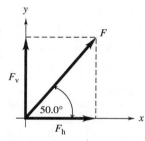

Figure 4.3

Solve: $F_h = F\cos 50.0° = 241$ N; $F_v = F\sin 50.0° = 287$ N.

Reflect: We could also find the components of $\vec{F}$ in the directions parallel and perpendicular to the incline but the problem asks for horizontal and vertical components.

4.5. Set Up: Let $F_1 = 985$ N, $F_2 = 788$ N, and $F_3 = 411$ N. The angles θ that each force makes with the $+x$-axis are $\theta_1 = 31°$, $\theta_2 = 122°$, and $\theta_3 = 233°$.

Solve: (a) $F_{1x} = F_1\cos\theta_1 = 844$ N; $F_{1y} = F_1\sin\theta_1 = 507$ N

$$F_{2x} = F_2\cos\theta_2 = -418 \text{ N}; F_{2y} = F_2\sin\theta_2 = 668 \text{ N}$$

$$F_{3x} = F_3\cos\theta_3 = -247 \text{ N}; F_{3y} = F_3\sin\theta_3 = -328 \text{ N}$$

(b) $R_x = F_{1x} + F_{2x} + F_{3x} = 179$ N; $R_y = F_{1y} + F_{2y} + F_{3y} = 847$ N

$$R = \sqrt{R_x^2 + R_y^2} = 886 \text{ N}; \tan\theta = \frac{R_y}{R_x} \text{ so } \theta = 78.1°$$

$\vec{R}$ and its components are shown in Figure 4.5.

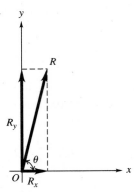

Figure 4.5

Reflect: Adding the forces as vectors gives a very different result from adding their magnitudes.

4.9. Set Up: Take $+x$ to be in the direction in which the cheetah moves. $v_{0x} = 0$.

Solve: (a) $v_x = v_{0x} + a_x t$ so $a_x = \dfrac{v_x - v_{0x}}{t} = \dfrac{20.1 \text{ m/s} - 0}{2.0 \text{ s}} = 10.05 \text{ m/s}^2$

$$F_x = ma_x = (68 \text{ kg})(10.05 \text{ m/s}^2) = 680 \text{ N}$$

(b) The force is exerted on the cheetah by the ground.

Reflect: The net force on the cheetah is in the same direction as the acceleration of the cheetah.

4.11. Set Up: Let $+x$ be the direction of the force. $\Sigma F_x = 80.0$ N. Use the information about the motion to find the acceleration and then use $\Sigma F_x = ma_x$ to calculate m.

Solve: $x - x_0 = 11.0$ m, $t = 5.00$ s, $v_{0x} = 0$. $x - x_0 = v_{0x}t + \frac{1}{2}a_x t^2$ gives

$$a_x = \frac{2(x - x_0)}{t^2} = \frac{2(11.0 \text{ m})}{(5.00 \text{ s})^2} = 0.880 \text{ m/s}^2.$$

$$m = \frac{\Sigma F_x}{a_x} = \frac{80.0 \text{ N}}{0.880 \text{ m/s}^2} = 90.9 \text{ kg}.$$

Reflect: The mass determines the amount of acceleration produced by a given force.

4.17. Set Up: $1 \text{ N} = 0.2248$ lb. $g = 9.80 \text{ m/s}^2 = 32.2 \text{ ft/s}^2$.

Solve: (a) A 1 kg stone has weight $w = mg = 9.8$ N; the 1 kg stone is heavier.

(b) A 1 slug stone has weight $w = mg = 32$ lb; the 1 slug stone is heavier.

(c) A 1 slug stone has weight $(32 \text{ lb})\left(\dfrac{1 \text{ N}}{0.2248 \text{ lb}}\right) = 140$ N; the 1 slug stone is heavier.

(d) A 1 kg stone weighs $w = mg = 9.8$ N. A 1 slug stone weighs 32 lb = 140 N. The 1 slug stone weighs more.

(e) A 1 kg stone weighs $w = mg = 9.8 \text{ N}\left(\dfrac{0.2248 \text{ lb}}{1 \text{ N}}\right) = 2.2$ lb. The 1 kg stone weighs more.

Reflect: kg and slug are units of mass; N and lb are units of force (weight).

4.19. Set Up: The weight of an object depends on its mass and the value of g at its location and is independent of the motion of the object.

Solve: (a) 138 N (b) 138 N (c) $w = mg = (138 \text{ kg})(9.80 \text{ m/s}^2) = 1350$ N, for both (a) and (b). (d) They would be the same, 138 N.

Reflect: The weight of an object is the gravitational force exerted on it by the earth.

4.21. Set Up: On Earth, $g_E = 9.80 \text{ m/s}^2$.

Solve: (a) $w_E = mg_E$ so $m = \dfrac{w_E}{g_E} = \dfrac{85.2 \text{ N}}{9.80 \text{ m/s}^2} = 8.69 \text{ kg}$. The mass m is independent of the location of the object.

$$w_M = mg_M \text{ so } g_M = \frac{w_M}{g} = \frac{32.2 \text{ N}}{8.69 \text{ kg}} = 3.71 \text{ m/s}^2.$$

(b) $m = 8.69 \text{ kg}$, the same at either location.

Reflect: The object weighs more on earth because the acceleration due to gravity is greater there.

4.23. Set Up: Use coordinates for which the $+x$-direction is the direction the car is traveling initially. $m = 1750 \text{ kg}$. When the car has stopped, $v_x = 0$.

Solve: (a) $f = 0.25mg$ and is in the $-x$-direction.

$$\Sigma F_x = ma_x \text{ gives } -f = ma_x \text{ and } a_x = \frac{-f}{m} = \frac{-0.25mg}{m} = -0.25g = -2.45 \text{ m/s}^2$$

The acceleration is 2.45 m/s^2, directed opposite to the motion.

(b) $v_{0x} = 110 \text{ km/h} = 30.6 \text{ m/s}$, $v_x = 0$, $a_x = -2.45 \text{ m/s}^2$. $v_x^2 = v_{0x}^2 + 2a_x(x - x_0)$ so

$$x - x_0 = \frac{v_x^2 - v_{0x}^2}{2a_x} = \frac{0 - (30.6 \text{ m/s})^2}{2(-2.45 \text{ m/s}^2)} = 191 \text{ m}$$

Reflect: The stopping distance is proportional to the square of the initial speed.

4.27. Set Up: The forces are gravity and contact forces from objects that touch the bottle. Neglect air resistance.

Solve: (a) (i) The forces are shown in Figure 4.27. P is the push, w is the gravity force, and n is the normal force applied by the table. n is perpendicular to the tabletop and f is parallel to the tabletop. (ii) The forces are the same as in (i), except now P is absent. (iii) The only force is the downward gravity force w.

Figure 4.27

(b) P, reaction force is the bottle pushing on the hand that pushes the bottle. n, reaction force is the bottle pushing down on the tabletop. w, reaction force is the bottle pulling upward on the earth.

Reflect: The reaction forces always are for a pair of objects.

4.29. Set Up: The brakes cause a friction force from the road that is directed opposite to the motion.

Solve: (a) The free-body diagram is shown in Figure 4.29a. w is the gravity force, n is the normal force and f is the friction force. The friction force is directed opposite to the direction the car is traveling.

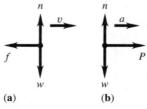

(a)   (b)

Figure 4.29

(b) The free-body diagram is shown in Figure 4.29b. The force P is applied to the passenger by the car seat and is in the direction the car is moving.

Reflect: In each case there is a net horizontal force on the object and this force produces an acceleration in the direction of the net force.

4.35. Set Up: Call the blocks A and B, with $w_A = 850$ N and $w_B = 750$ N. Since the worker pulls to the right, the friction force on each block is to the left. Let $\vec{F}$ be the force the worker applies and let T be the tension in the rope.
Solve: The free-body diagrams are shown in Figure 4.35. The rope pulls to the right on block A and to the left on block B.

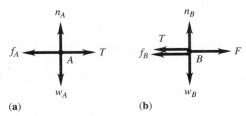

(a)   (b)

Figure 4.35

Reflect: The force $\vec{F}$ acts only on block B.

4.37. Set Up: Use coordinates with $+y$ upward. The objects are at rest so they have $a = 0$.
Solve: (a) The free-body diagram for the chandelier is shown in Figure 4.37a. T_b, the tension in the bottom link of the chain, equals the magnitude of the upward force that the bottom of the chain exerts on the chandelier. $\Sigma F_y = ma_y$ gives $T_b - w = 0$ so $T_b = w = 490$ N.

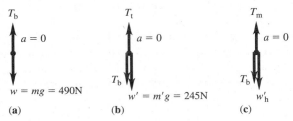

(a)   (b)   (c)

Figure 4.37

(b) Apply $\Sigma F_y = ma_y$ to the chain. The free-body diagram for the chain is shown in Figure 4.37b. The chandelier pulls downward on the chain with force T_b. T_t is the tension in the top link of the chain. $w' = 245$ N is the weight of the chain. We know from (a) that $T_b = 490$ N. $\Sigma F_y = ma_y$ gives $T_t - T_b - w' = 0$ and $T_t = T_b + w' = 735$ N.
(c) Apply $\Sigma F_y = ma_y$ to the bottom half of the chain. The free-body diagram is shown in Figure 4.37c. $w'_h = 122$ N is the weight of half the chain and T_m is the tension in the middle link of the chain. $\Sigma F_y = ma_y$ gives $T_m - T_b - w'_h = 0$ so $T_m = T_b + w'_h = 612$ N.
Reflect: The tension increases from the top to the bottom of the chain.

4.39. Set Up: Take $+y$ to be upward. The mass of the bucket is $m_B = w_B/g = 28.1$ kg and the mass of the chain is $m_C = w_C/g = 12.8$ kg. The chain also has an upward acceleration of 2.50 m/s².
Solve: (a) Consider the chain and bucket as a combined object of mass $m_{tot} = 28.1$ kg $+ 12.8$ kg $= 40.9$ kg and weight $w_{tot} = 400$ N. The free-body diagram is shown in Figure 4.39a. T_t is the tension in the top link. $\Sigma F_y = ma_y$ gives $T_t - w_{tot} = m_{tot}a$ and $T_t = w_{tot} + m_{tot}a = 400$ N $+ (40.9$ kg$)(2.50$ m/s²$) = 502$ N

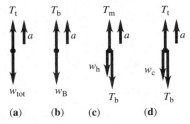

(a)   (b)   (c)   (d)

Figure 4.39

(b) Apply $\Sigma F_y = ma_y$ to the bucket. The free-body diagram (Figure 4.39b) gives $T_b - w_B = m_B a$ so $T_b = w_B + m_B a = 275$ N $+ (28.1$ kg$)(2.50$ m/s²$) = 345$ N

(c) Apply $\Sigma F_y = ma_y$ to the bottom half of the chain. The free-body diagram is shown in Figure 4.39c. The bottom half of the chain has mass $m_h = 6.4$ kg and weight $w_h = 62.5$ N. T_m is the tension in the middle link of the chain. $\Sigma F_y = ma_y$ gives $T_m - w_h - T_b = m_h a$.

$$T_m = w_h + T_b + m_h a = 62.5 \text{ N} + 345 \text{ N} + (6.4 \text{ kg})(2.5 \text{ m/s}^2) = 424 \text{ N}$$

Reflect: Part (b) can also be done by applying $\Sigma F_y = ma_y$ to the chain. The free-body diagram is given in Figure 4.39d. $\Sigma F_y = ma_y$ gives $T_t - w_c - T_b = m_c a$ and $T_b = T_t - w_c - m_c a = 502 \text{ N} - 125 \text{ N} - (12.8 \text{ kg})(2.50 \text{ m/s}^2) = 345$ N, which agrees with the value obtained previously. The tension increases from the bottom to the top of the chain.

4.43. Set Up: Your mass is $m = w/g = 63.8$ kg. Both you and the package have the same acceleration as the elevator. Take $+y$ to be upward, in the direction of the acceleration of the elevator.
Solve: **(a)** Your free-body diagram is shown in Figure 4.43a. n is the scale reading. $\Sigma F_y = ma_y$ gives $n - w = ma$ and $n = w + ma = 625 \text{ N} + (63.8 \text{ kg})(2.50 \text{ m/s}^2) = 784$ N

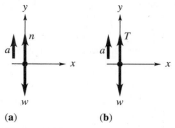

(a)   **(b)**

Figure 4.43

(b) The free-body diagram for the package is given in Figure 4.43b. $\Sigma F_y = ma_y$ gives $T - w = ma$ so $T = w + ma = (3.85 \text{ kg})(9.80 \text{ m/s}^2 + 2.50 \text{ m/s}^2) = 47.4$ N
Reflect: The objects accelerate upward so for each object the upward force is greater than the downward force.

4.45. Set Up: Take $+y$ to be upward. 4973 km/h $= 1381$ m/s. 45 km $= 4.5 \times 10^4$ m. The astronaut and the shuttle will have the same acceleration.
Solve: **(a)** $v_{0y} = 0$, $v_y = 1381$ m/s, $y - y_0 = 4.5 \times 10^4$ m. $v_y^2 = v_{0y}^2 + 2a_y(y - y_0)$ gives

$$a_y = \frac{v_y^2 - v_{0y}^2}{2(y - y_0)} = \frac{(1381 \text{ m/s})^2 - 0}{2(4.5 \times 10^4 \text{ m})} = 21.2 \text{ m/s}^2$$

(b) The free-body diagram for the astronaut is shown in Figure 4.45. $\Sigma F_y = ma_y$ gives $n - mg = ma$ and $n = m(g + a) = (55.0 \text{ kg})(9.80 \text{ m/s}^2 + 21.2 \text{ m/s}^2) = 1700$ N

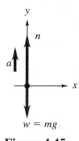

Figure 4.45

(c) She would think the scale reading is her weight, 1700 N. And she would therefore think her mass is $m = w/g = 173$ kg.
Reflect: Since she has an upward acceleration the upward force n is greater than the downward force of gravity.

4.47. Set Up: The car has mass $m = w/g = 1122$ kg. Use coordinates where the $+x$ axis is in the direction of the motion of the car. The friction force is then in the $-x$-direction. Since the car slows, its acceleration is opposite to is velocity.

Solve: **(a)** The free-body diagram for the car is given in Figure 4.47.

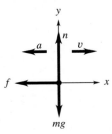

Figure 4.47

(b) The acceleration that slows the car is caused by the friction force.

(c) $x - x_0 = 44.5$ m, $v_{0x} = 28.7$ m/s, $v_x = 0$ (stops). $v_x^2 = v_{0x}^2 + 2a_x(x - x_0)$ gives

$$a_x = \frac{v_x^2 - v_{0x}^2}{2(x - x_0)} = \frac{0 - (28.7 \text{ m/s})^2}{2(44.5 \text{ m})} = -9.25 \text{ m/s}^2$$

$\Sigma F_x = ma_x$ gives $-f = ma_x$, so $f = -(1122 \text{ kg})(-9.25 \text{ m/s}^2) = 1.04 \times 10^4$ N

Reflect: The magnitude of the friction force between the tires and pavement determines the stopping distance.

4.49. **Set Up:** Take $+y$ to be upward.

Solve: **(a)** The free-body diagram for the froghopper while it is still pushing against the ground is given in Figure 4.49.

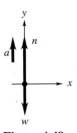

Figure 4.49

(b) $v_{0y} = 0$, $v_y = 4.0$ m/s, $t = 1.0 \times 10^{-3}$ s. $v_y = v_{0y} + a_y t$ gives

$$a_y = \frac{v_y - v_{0y}}{t} = \frac{4.0 \text{ m/s} - 0}{1.0 \times 10^{-3} \text{ s}} = 4.0 \times 10^3 \text{ m/s}^2.$$

$\Sigma F_y = ma_y$ gives $n - w = ma$, so $n = w + ma = m(g + a) = (12.3 \times 10^{-6} \text{ kg})(9.8 \text{ m/s}^2 + 4.0 \times 10^3 \text{ m/s}^2) = 0.049$ N

(c) $\dfrac{F}{w} = \dfrac{0.049 \text{ N}}{(12.3 \times 10^{-6} \text{ kg})(9.8 \text{ m/s}^2)} = 410$; $F = 410w$

Reflect: The ground pushes upward on the froghopper with a large force because the froghopper pushes downward on the ground with a large force.

4.51. **Set Up:** Let $+y$ be upward. At his maximum height, $v_y = 0$. While he is in the air, $a_y = -9.80$ m/s^2. $m = w/g = 90.8$ kg

Solve: **(a)** $v_{0y} = 0$, $a_y = -9.80$ m/s^2 and $y - y_0 = 1.2$ m. $v_y^2 = v_{0y}^2 + 2a_y(y - y_0)$ gives

$$v_{0y} = \sqrt{-2a_y(y - y_0)} = \sqrt{-2(-9.80 \text{ m/s}^2)(1.2 \text{ m})} = 4.85 \text{ m/s}$$

(b) For the motion while he is pushing against the floor, $v_{0y} = 0$, $v_y = 4.85$ m/s and $t = 0.300$ s. $v_y = v_{0y} + a_y$ gives

$$a_y = \frac{v_y - v_{0y}}{t} = \frac{4.85 \text{ m/s} - 0}{0.300 \text{ s}} = 16.2 \text{ m/s}^2. \text{ The acceleration is upward.}$$

(c) The free-body diagram while he is pushing against the floor is given in Figure 4.51. $\vec{F}$ is the vertical force the floor applies to him.

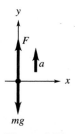

Figure 4.51

(d) $\Sigma F_y = ma_y$ gives $F - mg = ma$ and $F = m(g + a) = (90.8 \text{ kg})(9.80 \text{ m/s}^2 + 16.2 \text{ m/s}^2) = 2.36 \times 10^3 \text{ N}$. The force he applied to the ground has this same magnitude and is downward.

Reflect: The ground must push upward on him with a force greater than his weight in order to give him an upward acceleration.

4.57. Set Up: Let $+y$ be upward. $\vec{F}_{air}$ is the force of air resistance.

Solve: (a) $w = mg = (55.0 \text{ kg})(9.80 \text{ m/s}^2) = 539 \text{ N}$

(b) The free-body diagram is given in Figure 4.57. $\Sigma F_y = F_{air} - w = 620 \text{ N} - 539 \text{ N} = 81 \text{ N}$. The net force is upward.

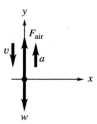

Figure 4.57

(c) $a_y = \dfrac{\Sigma F_y}{m} = \dfrac{81 \text{ N}}{55.0 \text{ kg}} = 1.5 \text{ m/s}^2$, upward.

Reflect: Both the net force and the acceleration are upward. Since her velocity is downward and her acceleration is upward, her speed decreases.

APPLICATIONS OF NEWTON'S LAWS

Problems 1, 3, 5, 9, 11, 19, 21, 25, 29, 31, 33, 37, 39, 41, 45, 47, 51, 55, 59, 61, 63, 65, 67, 71, 75, 79, 81, 85, 87

Solutions to Problems

5.1. Set Up: Constant speed means $a = 0$. Use coordinates where $+y$ is upward. The breaking strength for the rope in each part is (a) 50 N, (b) 4000 N, and (c) 6×10^4 N. The free-body diagram for the bucket plus its contents is given in Figure 5.1.

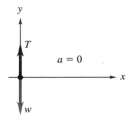

Figure 5.1

Solve: $\Sigma F_y = ma_y$ gives $T - w = 0$ so $T = w$.
(a) $w = 50$ N, so 35 N of cement. The mass of cement is 3.6 kg.
(b) $w = 4000$ N, so 3985 N of cement. The mass of cement is 410 kg.
(c) $w = 6 \times 10^4$ N, so 6×10^4 N of cement. The mass of cement is 6×10^3 kg.
Reflect: Since $a = 0$, the upward pull of the rope equals the total weight of the object.

5.3. Set Up: $a = 0$ for each object. Apply $\Sigma F_y = ma_y$ to each weight and to the pulley. Take $+y$ upward. The pulley has negligible mass. Let T_r be the tension in the rope and let T_c be the tension in the chain.
Solve: (a) The free-body diagram for each weight is the same and is given in Figure 5.3a. $\Sigma F_y = ma_y$ gives $T_r = w = 25.0$ N.

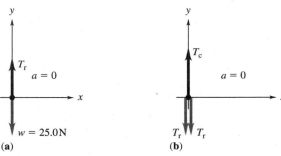

Figure 5.3

(b) The free-body diagram for the pulley is given in Figure 5.3b. $T_c = 2T_r = 50.0$ N.
Reflect: The tension is the same at all points along the rope.

5.5. Set Up: Take $+y$ upward. Apply $\Sigma\vec{F} = m\vec{a}$ to the person. $a = 0$.

Solve: (a) The free-body diagram for the person is given in Figure 5.5. The two sides of the rope each exert a force with vertical component $T\sin\theta$. $\Sigma F_y = ma_y$ gives $2T\sin\theta - w = 0$.

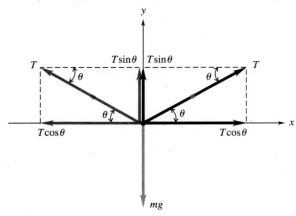

Figure 5.5

$$T = \frac{mg}{2\sin\theta} = \frac{(90.0\ \text{kg})(9.80\ \text{m/s}^2)}{2\sin 10.0°} = 2.54 \times 10^3\ \text{N}$$

(b) Set $T = 2.50 \times 10^4$ N and solve for θ:

$$\sin\theta = \frac{mg}{2T} = \frac{(90.0\ \text{kg})(9.80\ \text{m/s}^2)}{2(2.50 \times 10^4\ \text{N})} \text{ and } \theta = 1.01°.$$

Reflect: Only the vertical components of the tension on each side of the person act to hold him up. The tension in the rope is much greater than his weight.

5.9. Set Up: Use coordinates where $+y$ is upward and $+x$ is to the right. The force at the foot is horizontal. The tension T in the cable that has one end attached to the leg and the other end attached to the weight is equal to W and is the same everywhere along the cable.

Solve: (a) The free-body diagram for the leg is given in Figure 5.9a. F_T is the traction force on the leg. $\sum F_y = 0$ gives $T - w = 0$ and $T = w = 47.0$ N so $W = 47.0$ N.

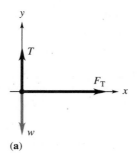

(a)

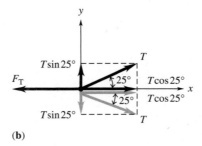

(b)

Figure 5.9

(b) The free-body diagram for the pulley that is attached to the foot is given in Figure 5.9b. $\sum F_x = 0$ gives $2T\cos 25° - F_T = 0$ so $F_T = 2T\cos 25° = 85.2$ N.
Reflect: The tractive force on the leg is greater than W.

5.11. Set Up: Use coordinates with $+y$ upward and $+x$ to the right. Call the objects A and B, with $w_A = 175$ N and $m_B = 32$ kg. Label the tensions in the three wires T_1, T_2, and T_3.
Solve: (a) The free-body diagrams for each object are given in Figure 5.11. T_3 has been replaced by its x and y components.

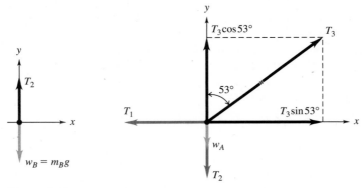

Figure 5.11

(b) $\sum F_y = 0$ for B gives $T_2 - w_B = 0$ and $T_2 = m_B g = 314$ N. $\sum F_y = 0$ for A gives $T_3\cos 53° - T_2 - w_A = 0$ so
$T_3 = \dfrac{T_2 + w_A}{\cos 53°} = 813$ N. $\sum F_x = 0$ for A gives $T_3\sin 53° - T_1 = 0$ and $T_1 = 649$ N.
(c) The tensions are unaffected by the length of the wires, so long as the third wire still makes a 53° angle with one wall.
Reflect: $T_3\cos 53°$ equals the combined weight of both objects, 489 N.

5.19. Set Up: Solve for the net force F_{net} on the ball.

Solve: $v_0 = 0$, $v = 73.14$ m/s, $t = 30 \times 10^{-3}$ s. $v = v_0 + at$ gives

$$a = \frac{v - v_0}{t} = \frac{73.14 \text{ m/s}}{30 \times 10^{-3} \text{ s}} = 2.44 \times 10^3 \text{ m/s}^2.$$

$$F_{net} = ma = (W/g)a = 250W.$$

Reflect: The acceleration is much larger than g and the net force is much larger than the weight of the ball.

5.21. Set Up: Take $+y$ to be upward. After he leaves the ground the person travels upward 60 cm and his acceleration is $g = 9.80$ m/s^2, downward. His weight is W so his mass is W/g.

Solve: (a) $v_y = 0$ (at the maximum height), $y - y_0 = 0.60$ m, $a_y = -9.80$ m/s^2. $v_y^2 = v_{0y}^2 + 2a_y(y - y_0)$ gives

$$v_{0y} = \sqrt{-2a_y(y - y_0)} = \sqrt{-2(-9.80 \text{ m/s}^2)(0.60 \text{ m})} = 3.4 \text{ m/s}.$$

(b) The free-body diagram for the person while he is pushing up against the ground is given in Figure 5.21.

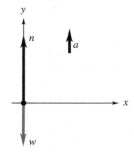

Figure 5.21

(c) For the jump, $v_{0y} = 0$, $v_y = 3.4$ m/s (from part (a)), $y - y_0 = 0.50$ m. $v_y^2 = v_{0y}^2 + 2a_y(y - y_0)$ gives

$$a_y = \frac{v_y^2 - v_{0y}^2}{2(y - y_0)} = \frac{(3.4 \text{ m/s})^2 - 0}{2(0.50 \text{ m})} = 11.6 \text{ m/s}^2$$

$$\Sigma F_y = ma_y \text{ gives } n - W = ma. \; n = W + ma = W\left(1 + \frac{a}{g}\right) = 2.2W$$

Reflect: To accelerate the person upward during the jump, the upward force from the ground must exceed the downward pull of gravity. The ground pushes up on him because he pushes down on the ground.

5.25. Set Up: The camera is at rest relative to the train car and has the same horizontal acceleration as the car. Use coordinates with $+y$ upward and $+x$ in the direction of motion of the train car.

Solve: **(a)** The free-body-diagram is given in Figure 5.25. The tension T in the strap has been replaced by its x and y components. $\phi = 12.0°$

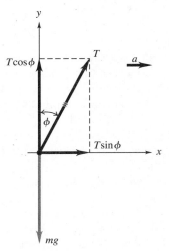

Figure 5.25

(b) $\Sigma F_x = ma_x$ gives $T\sin\phi = ma$. $\Sigma F_y = ma_y$ gives $T\cos\phi - mg = 0$. Combining these two equations to eliminate T gives $a = g\tan\phi = 2.08$ m/s^2. And the y component equation gives

$$T = \frac{mg}{\cos\phi} = \frac{3.00 \text{ N}}{\cos 12.0°} = 3.07 \text{ N}$$

Reflect: As a increases there must be a larger horizontal component of T and the angle ϕ increases.

5.29. Set Up: Let $m_1 = 50.0$ kg and $m_2 = 30.0$ kg. The two boxes have the same magnitude of acceleration. For the 30.0 kg box let $+y$ be downward and for the 50.0 kg box use coordinates parallel and perpendicular to the ramp, with $+x$ up the ramp.
Solve: **(a)** The free-body diagrams are given in Figure 5.29. $\phi = 30.0°$

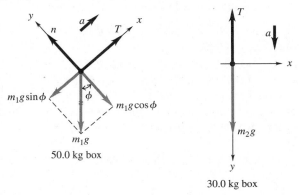

Figure 5.29

(b) The downward force on the 30.0 kg box is $m_2 g = 294$ N. The force pulling the system in the opposite direction is the component of $m_1 g$ directed down the ramp, $m_1 g\sin\phi = 245$ N. $m_2 g > m_1 g\sin\phi$ and the 30.0 kg box moves downward. Therefore, the 50.0 kg box moves up the ramp.
(c) $\Sigma F_x = ma_x$ applied to m_1 gives $T - m_1 g\sin\phi = m_1 a$. $\Sigma F_y = ma_y$ applied to m_2 gives $m_2 g - T = m_2 a$. Combining these two equations to eliminate T gives

$$a = \frac{m_2 g - m_1 g\sin\phi}{m_1 + m_2} = \frac{294 \text{ N} - 245 \text{ N}}{50.0 \text{ kg} + 30.0 \text{ kg}} = 0.612 \text{ m/s}^2$$

The 50.0 kg box accelerates up the ramp at 0.612 m/s^2 and the 30.0 kg box accelerates downward at 0.612 m/s^2.

Reflect: Only the component of the weight of the 50 kg box that is parallel to the ramp acts to oppose the weight of the 30 kg box. Therefore, the 30 kg box pulls the 50 kg box up the ramp even though its weight is less than the weight of the 50 kg box.

5.31. Set Up: Since the only vertical forces are n and w, the normal force on the box equals its weight. Static friction is as large as it needs to be to prevent relative motion between the box and the surface, up to its maximum possible value of $f_s^{max} = \mu_s n$. If the box is sliding then the friction force is $f_k = \mu_k n$.
Solve: **(a)** If there is no applied force, no friction force is needed to keep the box at rest.
(b) $f_s^{max} = \mu_s n = (0.40)(40.0 \text{ N}) = 16.0 \text{ N}$. If a horizontal force of 6.0 N is applied to the box, then $f_s = 6.0 \text{ N}$ in the opposite direction.
(c) The monkey must apply a force equal to f_s^{max}, 16.0 N.
(d) Once the box has started moving, a force equal to $f_k = \mu_k n = 8.0 \text{ N}$ is required to keep it moving at constant velocity.
Reflect: $\mu_k < \mu_s$ and less force must be applied to the box to maintain its motion than to start it moving.

5.33. Set Up: The free-body diagram for the two crates treated as a single object, weight w_C, is shown in Figure 5.33a. The system doesn't move so the friction force exerted by the roof is static friction. For the heaviest pallet of bricks this force has its maximum possible, $f_s = \mu_s n$. The free-body diagram for the pallet of bricks is given in Figure 5.33b.

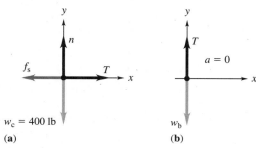

Figure 5.33

Solve: **(a)** For the crates, $\Sigma F_y = ma_y$ gives $n - w_C = 0$ and $n = 400$ lb. Then
$$f_s = \mu_s n = (0.666)(400 \text{ lb}) = 266 \text{ lb}.$$
$\Sigma F_x = ma_x$ gives $T - f_s = 0$ and $T = f_s = 266$ lb
For the bricks, $\Sigma F_y = ma_y$ gives $T - w_b = 0$ and $w_b = T = 266$ lb
(b) For the upper crate the only horizontal force on the crate would be friction. This crate has $a_x = 0$ so $\Sigma F_x = 0$ and the friction force is zero.
Reflect: If some bricks are removed, so the weight of the pallet is reduced, the system remains at rest. The friction force on the crate is equal to the new weight of the pallet and is less than $\mu_s n$.

5.37. Set Up: The free-body diagram for the car is shown in Figure 5.37.

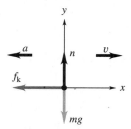

Figure 5.37

Solve: Relate a to μ_k. $\Sigma F_y = ma_y$ gives $n = mg$. $\Sigma F_x = ma_x$ gives $-f_k = -ma$. $\mu_k mg = ma$ and $a = \mu_k g$. Use a kinematic equation to relate a to the stopping distance d. $a_x = -a$, $v_x = 0$ (stops), $x - x_0 = d$. $v_x^2 = v_{0x}^2 + 2a_x(x - x_0)$ gives $v_0^2 = +2ad$. Same v_0 means $a_d d_d = a_w d_w$ and $\mu_{kd} d_d = \mu_{kw} d_w$.

$$d_w = \left(\frac{\mu_{kd}}{\mu_{kw}}\right) d_d.$$

$d_d = D$ and $\mu_{kw} = \frac{1}{8}\mu_{kd}$ so $d_w = 8D$.

Reflect: The stopping distance is proportional to the acceleration and the acceleration is proportional to the friction force, so the stopping distance is decreased by the same factor as is the coefficient of friction.

5.39. Set Up: $n = mg$. For constant speed, $a_x = 0$. Apply $\Sigma F_x = ma_x$ with $+x$ in the direction of motion of the box.
Solve: (a) The free-body diagram for the box is given in Figure 5.39. $\Sigma F_x = ma_x$ gives $P - f_k = 0$ and $P = f_k = \mu_k mg = (0.200)(11.2 \text{ kg})(9.80 \text{ m/s}^2) = 22.0$ N.

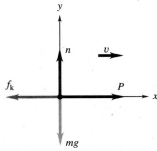

Figure 5.39

(b) $P = 0$ and $-f_k = ma_x$. $a_x = -\dfrac{f_k}{m} = -\mu_k g = -1.96 \text{ m/s}^2$.

Reflect: In (b) the friction force and the acceleration are directed opposite to the motion. Since $\vec{a}$ and $\vec{v}$ are in opposite directions, the box slows down.

5.41. Set Up: The free-body diagram for the box is given in Figure 5.41. When the box is either at rest or sliding at constant speed, the acceleration of the box is zero. Use coordinates as shown, with the x axis parallel to the incline. The weight mg has been resolved into its x and y components.

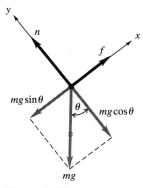

Figure 5.41

Solve: $\Sigma F_y = ma_y$ so $n - mg\cos\theta = 0$ and $n = mg\cos\theta$. $\Sigma F_x = ma_x$ so $f - mg\sin\theta = 0$
$f = \mu n = \mu mg\cos\theta$ so $\mu mg\cos\theta = mg\sin\theta$ and $\mu = \tan\theta$. At $\theta = \theta_1$, $f = f_s$ and $\mu_s = \tan\theta_1$. At $\theta = \theta_2$, $f = f_k$ and $\mu_k = \tan\theta_2$.
Reflect: Usually $\mu_k < \mu_s$, so $\theta_2 < \theta_1$. If the angle is kept at θ_1 the box will accelerate down the incline once it has been set into motion.

5.45. Set Up: The free-body diagram for the crate is given in Figure 5.45. Use coordinates with axes parallel and perpendicular to the ramp. Constant speed means $a = 0$. The mass of the crate is $m = w/g = 38.3$ kg. The weight mg has been resolved into its x and y components.

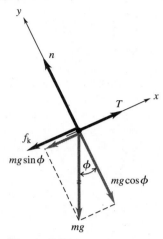

Figure 5.45

Solve: (a) $\Sigma F_y = ma_y$ gives $n = mg\cos\phi$. $f_k = \mu_k n = \mu_k mg\cos\phi$. $\Sigma F_x = ma_x$ with $a_x = 0$ gives $T - mg\sin\phi - f_k = 0$, so

$$T = mg\sin\phi + f_k = (375\text{ N})\sin 33° + (0.250)(375\text{ N})\cos 33° = 283\text{ N}$$

(b) $T = 0$ so $-f_k - mg\sin\phi = ma_x$

$$-\mu_k mg\cos\phi - mg\sin\phi = ma_x \text{ and } a_x = -g(\mu_k\cos\phi + \sin\phi) = -0.754g$$

The acceleration would be $0.754g = 7.39$ m/s^2, directed down the ramp.

Reflect: In part (a) the force up the incline is 283 N and the force down the incline is 283 N. When the rope breaks $T \to 0$ and the force down the incline remains 283 N. The acceleration down the incline is $(283\text{ N})/(38.3\text{ kg}) = 7.39$ m/s^2.

5.47. Set Up: The free-body diagram for the load is given in Figure 5.47. The tension T in the rope has been resolved into its x and y components.

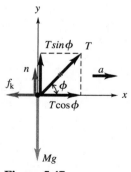

Figure 5.47

Solve: (a) $\Sigma F_y = ma_y$ gives $T\sin\phi + n - Mg = 0$ and $n = Mg - T\sin\phi$.

$$f = \mu n = \mu Mg - \mu T\sin\phi$$

$\Sigma F_x = ma_x$ gives $T\cos\phi - f = Ma$ so $T\cos\phi - \mu Mg + \mu T\sin\phi = Ma$.

$$T = \frac{M(a + \mu g)}{\cos\phi + \mu\sin\phi}$$

(b) (i) No friction means $\mu = 0$. The result in (a) becomes $T\cos\phi = Ma$.

(ii) Pulls horizontally means $\phi = 0°$. The result in (a) becomes $T = M(a + \mu g)$.

Reflect: Pulling above the horizontal reduces the normal force and therefore reduces the friction force. But it also reduces the component of T in the direction of the acceleration. As ϕ is increased, T decreases, reaches a minimum value at an optimum value of ϕ and then increases. The optimum angle depends on the value of μ.

5.51. Set Up: $F_D = Dv^2$

Solve: **(a)** At the terminal speed v_t, $F_D = W$.

(b) $W = Dv_t^2$. When $v = v_t/2$, $F_D = D(v_t/2)^2 = \frac{1}{4}Dv_t^2 = W/4$

(c) When $v = v_t/3$, $F_D = W/9$.

Reflect: At the terminal velocity the upward drag force equals the downward gravity force and the acceleration is zero. At speeds less than v_t the upward drag force is less than the weight and the person accelerates downward.

5.55. Set Up: $|F_{spr}| = mg$ so $mg = kx$. x is the change in length of the spring.

Solve: **(a)** $k = \dfrac{mg}{x} = \dfrac{(0.875\ \text{kg})(9.80\ \text{m/s}^2)}{0.0240\ \text{m}} = 357\ \text{N/m}$

(b) $m = \dfrac{kx}{g} = \dfrac{(357\ \text{N/m})(0.0572\ \text{m})}{9.80\ \text{m/s}^2} = 2.08\ \text{kg}$

Reflect: The elongation of the vertical spring is proportional to the mass hung from it. So, the mass in (b) is

$$\left(\frac{5.72\ \text{cm}}{2.40\ \text{cm}}\right)(0.875\ \text{kg}) = 2.08\ \text{kg}.$$

5.59. Set Up: $mg = |F_{spr}| = kx$

Solve: **(a)** Yes. Hooke's law says $m = \left(\dfrac{k}{g}\right)x$ and this is true here since the graph of m versus x is a straight line.

(b) The slope of the graph of m versus x is k/g, so $\dfrac{k}{g} = \dfrac{0.500\ \text{kg}}{0.200\ \text{m}} = 2.50\ \text{kg/m}$ and $k = (2.50\ \text{kg/m})g = 24.5\ \text{N/m}$

(c) $x = \dfrac{mg}{k} = \dfrac{(2.35\ \text{kg})(9.80\ \text{m/s}^2)}{24.5\ \text{N/m}} = 0.94\ \text{m} = 94\ \text{cm}$

Reflect: A 0.50 kg mass stretches the spring 20 cm, so a 2.35 kg mass stretches it $\left(\dfrac{2.35\ \text{kg}}{0.50\ \text{kg}}\right)(20\ \text{cm})$.

5.61. Set Up: The magnitude of the spring force is kx. The force diagrams are given in Figure 5.61.

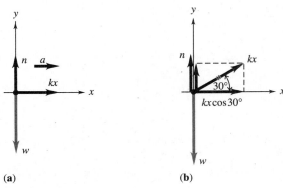

(a)   (b)

Figure 5.61

Solve: **(a)** $kx = ma$ and $x = \dfrac{ma}{k} = \dfrac{(9.50\ \text{kg})(2.00\ \text{m/s}^2)}{125\ \text{N/m}} = 0.152\ \text{m} = 15.2\ \text{cm}$

(b) $kx\cos 30.0° = ma$ and $x = \dfrac{ma}{k\cos 30.0°} = \dfrac{0.152\ \text{m}}{\cos 30.0°} = 0.176\ \text{m} = 17.6\ \text{cm}$

Reflect: Only the horizontal component of the spring force accelerates the sled, so a greater spring force is needed in part (b).

5.63. Set Up: Take $+y$ downward. (a) Assume the hip is in free-fall. (b) The free-body diagram for the person is given in Figure 5.63. It is assumed that the whole mass of the person has the same acceleration as her hip.

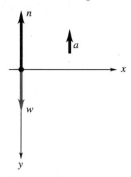

Figure 5.63

Solve: (a) $v_{0y} = 0$, $y - y_0 = 1.0$ m, $a_y = +9.80$ m/s^2. $v_y^2 = v_{0y}^2 + 2a_y(y - y_0)$ gives

$$v_y = \sqrt{2a_y(y - y_0)} = \sqrt{2(9.80 \text{ m/s}^2)(1.0 \text{ m})} = 4.4 \text{ m/s}$$

(b) $v_{0y} = 4.4$ m/s, $y - y_0 = 0.020$ m, $v_y = 1.3$ m/s. $v_y^2 = v_{0y}^2 + 2a_y(y - y_0)$ gives

$$a_y = \frac{v_y^2 - v_{0y}^2}{2(y - y_0)} = \frac{(1.3 \text{ m/s})^2 - (4.4 \text{ m/s})^2}{2(0.020 \text{ m})} = -440 \text{ m/s}^2$$

The acceleration is 440 m/s^2, upward. $\Sigma F_y = ma_y$ gives $w - n = -ma$ and

$$n = w + ma = m(a + g) = (55 \text{ kg})(440 \text{ m/s}^2 + 9.80 \text{ m/s}^2) = 25{,}000 \text{ N}$$

(c) $v_y = v_{0y} + a_y t$ gives $t = \dfrac{v_y - v_{0y}}{a_y} = \dfrac{1.3 \text{ m/s} - 4.4 \text{ m/s}}{-440 \text{ m/s}^2} = 7.0$ ms

Reflect: When the velocity change occurs over a small distance the acceleration is large.

5.65. Set Up: For each spring $F = kx$, so $F_1 = k_1 x$ and $F_2 = k_2 x$.
Solve: (a) The end of each spring moves the same distance, so $x_1 = x_2 = x$.
(b) The free-body diagram for the point where F is applied is given in Figure 5.65.

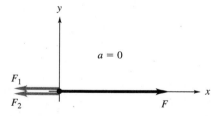

Figure 5.65

$$F = F_1 + F_2 = k_1 x + k_2 x = (k_1 + k_2)x$$

(c) For the single effective spring $F = k_{\text{eff}} x$. Comparing this to the result in (b) gives $k_{\text{eff}} = k_1 + k_2$.
(d) $k_{\text{eff}} = k_1 + k_2 = 25 \text{ N/cm} + 45 \text{ N/cm} = 70 \text{ N/cm}$
Reflect: For two springs in parallel the effective force constant for the combination is greater than the force constant of either spring.

5.67. Set Up: For the climber, $a = 0$. The rock face exerts a horizontal force n on the climber and no friction force. Let $+y$ be upward.

Solve: The free-body diagram for the climber is given in Figure 5.67.

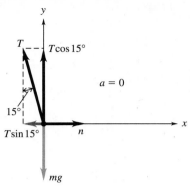

Figure 5.67

$\Sigma F_y = ma_y$ gives $T\cos 15° - mg = 0$.

$$T = \frac{mg}{\cos 15°} = \frac{600.0 \text{ N}}{\cos 15°} = 621 \text{ N}.$$

(b) $\Sigma F_x = ma_x$ gives $n - T\sin 15° = 0$. $n = T\sin 15° = 161$ N.

Reflect: The tension in the rope is greater than his weight because only the vertical component of the tension holds him up.

5.71. Set Up: $a = 0$ for both objects. The tension in the vertical wire is w. Apply $\Sigma\vec{F} = m\vec{a}$ to the knot where the wires are joined and then to block A. The static friction force on block A will be whatever is required to keep block A at rest.

Solve: The free-body diagram for the knot is given in Figure 5.71a. $\Sigma F_y = ma_y$ gives $T_1\sin 45° = w$. $\Sigma F_x = ma_x$ gives $T_2\cos 45.0° = T_3$. $\sin 45.0° = \cos 45.0°$ and $T_3 = w$. The free-body diagram for block A is given in Figure 5.71b. $f_s = T_3 = w = 12.0$ N

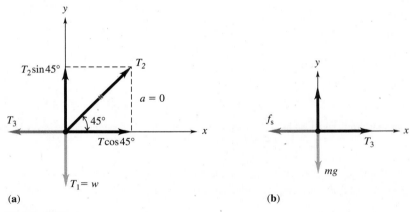

(a)   (b)

Figure 5.71

(b) $f_s^{\text{max}} = \mu_s n = \mu_s mg = (0.25)(60.0 \text{ N}) = 15.0$ N. Then $w = f_s^{\text{max}} = 15.0$ N.

Reflect: The friction force found in part (a) is less than the maximum possible friction force.

5.75. Set Up: On Mars, $g_M = 0.379g_E = 3.71 \text{ m/s}^2$. The free-body diagrams in each stage are shown in Figure 5.75a-c. F_D is the air drag force in Stage I, F_P is the force from the parachute in Stage II and F_R is the force from the retrorockets in Stage III.

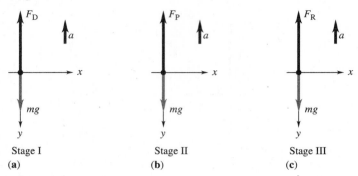

Stage I   Stage II   Stage III
(a)   **(b)**   **(c)**

Figure 5.75

Solve: (a) $v_{0y} = 19{,}300 \text{ km/h} = 5361 \text{ m/s}$, $v_y = 1600 \text{ km/h} = 444 \text{ m/s}$, $t = 240 \text{ s}$. $v_y = v_{0y} + a_y t$ gives

$$a_y = \frac{v_y - v_{0y}}{t} = \frac{444 \text{ m/s} - 5361 \text{ m/s}}{240 \text{ s}} = -20.5 \text{ m/s}^2.$$

$\Sigma F_y = ma_y$ gives $mg - F_D = ma_y$.

$$F_D = m(g - a_y) = (827 \text{ kg})(3.71 \text{ m/s}^2 - [-20.5 \text{ m/s}^2]) = 2.0 \times 10^4 \text{ N}$$

(b) $v_{0y} = 1600 \text{ km/h} = 444 \text{ m/s}$, $v_y = 321 \text{ km/h} = 89.2 \text{ m/s}$, $t = 94 \text{ s}$. $v_y = v_{0y} + a_y t$ gives

$$a_y = \frac{v_y - v_{0y}}{t} = \frac{89.2 \text{ m/s} - 444 \text{ m/s}}{94 \text{ s}} = -3.77 \text{ m/s}^2.$$

$\Sigma F_y = ma_y$ gives $mg - F_P = ma_y$.

$$F_P = m(g - a_y) = (827 \text{ kg})(3.71 \text{ m/s}^2 - [-3.77 \text{ m/s}^2]) = 6.19 \times 10^3 \text{ N}$$

(c) $v_{0y} = 321 \text{ km/h} = 89.2 \text{ m/s}$, $v_y = 0$, $t = 2.5 \text{ s}$. $v_y = v_{0y} + a_y t$ gives

$$a_y = \frac{v_y - v_{0y}}{t} = \frac{0 - 89.2 \text{ m/s}}{2.5 \text{ s}} = -35.7 \text{ m/s}^2.$$

$\Sigma F_y = ma_y$ gives $mg - F_R = ma_y$.

$$F_R = m(g - a_y) = (827 \text{ kg})(3.71 \text{ m/s}^2 - [-35.7 \text{ m/s}^2]) = 3.26 \times 10^4 \text{ N}$$

Reflect: In each stage the upward force is greater than the downward force, the acceleration is upward, opposite to $\vec{v}$, and the speed decreases.

5.79. Set Up: If the truck accelerates to the right the friction force on the box is to the right, to try to prevent the box from sliding relative to the truck. The free-body diagram for the box is given in Figure 5.79. The maximum acceleration of the box occurs when f_s has its maximum value, $f_s = \mu_s n$. If the box doesn't slide its acceleration equals the acceleration of the truck.

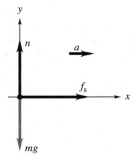

Figure 5.79

Solve: $n = mg$. $\Sigma F_x = ma_x$ gives $f_s = ma$ so $\mu_s mg = ma$ and $a = \mu_s g = 6.37 \text{ m/s}^2$. $v_{0x} = 0$, $v_x = 30.0 \text{ m/s}$. $v_x = v_{0x} + a_x t$ gives

$$t = \frac{v_x - v_{0x}}{a_x} = \frac{30.0 \text{ m/s} - 0}{6.37 \text{ m/s}^2} = 4.71 \text{ s}$$

Reflect: If the truck has a smaller acceleration it is still true that $f_s = ma$, but now $f_s < \mu_s n$.

5.81. Set Up: Write the air resistance force as $F_D = Dv^2$. k has units of kg/m.
Solve: (a) At the terminal velocity, $F_D = mg$ and $Dv_t^2 = mg$

without a parachute: $D_{wo} = \dfrac{mg}{v_t^2} = \dfrac{(75 \text{ kg})(9.80 \text{ m/s}^2)}{(60 \text{ m/s})^2} = 0.20 \text{ kg/m}$

with a parachute: $D_w = \dfrac{(75 \text{ kg})(9.80 \text{ m/s}^2)}{(6 \text{ m/s})^2} = 20 \text{ kg/m}$

(b) $v_t = \sqrt{\dfrac{mg}{D}}$, so $\dfrac{v_t}{\sqrt{m}}$ is constant. $\dfrac{v_{t1}}{\sqrt{m_1}} = \dfrac{v_{t2}}{\sqrt{m_2}}$ and $v_{t2} = v_{t1}\sqrt{\dfrac{m_2}{m_1}}$.

without: $v_{t2} = (60 \text{ m/s})\sqrt{\dfrac{62 \text{ kg}}{75 \text{ kg}}} = 55 \text{ m/s}$ and *with:* $v_{t2} = (6 \text{ m/s})\sqrt{\dfrac{62 \text{ kg}}{75 \text{ kg}}} = 5.5 \text{ m/s}$

Reflect: At the terminal velocity the air resistance force equals the weight, so a lighter person has a smaller terminal velocity.

5.85. Set Up: The free-body diagrams for the boxes are shown in Figure 5.85. Label the boxes A and B, with $m_A = 2.0 \text{ kg}$ and $m_B = 3.0 \text{ kg}$. F is the spring force and is the same for each box.

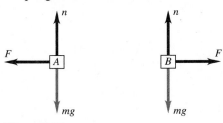

Figure 5.85

Solve: $F = k|x| = (250 \text{ N/m})(0.060 \text{ m}) = 15.0 \text{ N}$

$$a_A = \frac{F}{m_A} = \frac{15.0 \text{ N}}{2.0 \text{ kg}} = 7.5 \text{ m/s}^2;$$

$$a_B = \frac{F}{m_B} = \frac{15.0 \text{ N}}{3.0 \text{ kg}} = 5.0 \text{ m/s}^2$$

The accelerations are in opposite directions.

Reflect: The same magnitude of force is exerted on each object, but the acceleration that is produced by this force is larger for the object of smaller mass.

5.87. Set Up: The block has the same horizontal acceleration a as the cart. Let $+x$ be to the right and $+y$ be upward. To find the minimum acceleration required, set the static friction force equal to its maximum value, $\mu_s n$.
Solve: The free-body diagram for the block is given in Figure 5.87.

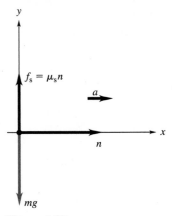

Figure 5.87

$\Sigma F_x = ma_x$ gives $n = ma$. $\Sigma F_y = ma_y$ gives $f_s - mg = 0$. $\mu_s n = mg$ and $\mu_s ma = mg$. $a = \dfrac{g}{\mu_s}$.

Reflect: The smaller μ_s is the greater a must be to prevent slipping. Increasing a increases the normal force n and that increases the maximum f_s for a given μ_s.

6

CIRCULAR MOTION AND GRAVITATION

Problems 1, 3, 5, 9, 11, 13, 15, 19, 21, 23, 27, 29, 37, 41, 47, 49, 53, 55, 57, 59, 61

Solutions to Problems

6.1. Set Up: The path shows the relative size of the radius of curvature R of each part of the track. $R_B > R_C$. A straight line is an arc of a circle of infinite radius, so $R_A \rightarrow \infty$.

Solve: $a_{\text{rad}} = \dfrac{v^2}{R}$. At A, $a_{\text{rad}} = 0$. $a_{\text{rad},B} < a_{\text{rad},C}$. In each case, $\vec{a}_{\text{rad}}$ is directed toward the center of the circular path. The net force is proportional to $\vec{a}_{\text{rad}}$ and is shown in Figure 6.1.

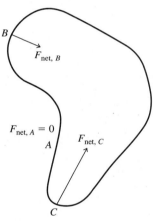

Figure 6.1

Reflect: The road must exert a greater friction force on the car when it makes tighter turns that have smaller radii of curvature.

6.3. Set Up: Each hand travels in a circle of radius 0.750 m and has mass $(0.0125)(52 \text{ kg}) = 0.65 \text{ kg}$ and weight 6.4 N. The period for each hand is $T = (1.0 \text{ s})/(2.0) = 0.50 \text{ s}$. Let $+x$ be toward the center of the circular path.
(a) The free-body diagram for one hand is given in Figure 6.3. $\vec{F}$ is the force exerted on the hand by the wrist. This force has both horizontal and vertical components.

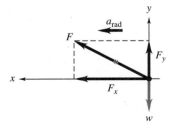

Figure 6.3

(b) $a_{rad} = \dfrac{4\pi^2 R}{T^2} = \dfrac{4\pi^2(0.750 \text{ m})}{(0.50 \text{ s})^2} = 118 \text{ m/s}^2$

$\Sigma F_x = ma_x$ gives $F_x = ma_{rad} = (0.65 \text{ kg})(118 \text{ m/s}^2) = 77 \text{ N}$

(c) $\dfrac{F}{w} = \dfrac{77 \text{ N}}{6.4 \text{ N}} = 12$. The horizontal force from the wrist is 12 times the weight of the hand.

Reflect: The wrist must also exert a vertical force on the hand equal to its weight.

6.5. Set Up: The person moves in a circle of radius $R = 3.00 \text{ m} + (5.00 \text{ m})\sin 30.0° = 5.50 \text{ m}$. The acceleration of the person is $a_{rad} = v^2/R$, directed horizontally to the left in the figure in the problem. The time for one revolution is the period $T = \dfrac{2\pi R}{v}$.

Solve: (a) The free-body diagram is given in Figure 6.5. $\vec{F}$ is the force applied to the seat by the rod.

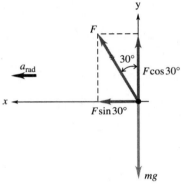

Figure 6.5

(b) $\Sigma F_y = ma_y$ gives $F\cos 30.0° = mg$ and $F = \dfrac{mg}{\cos 30.0°}$. $\Sigma F_x = ma_x$ gives $F\sin 30.0° = m\dfrac{v^2}{R}$. Combining these two equations gives

$$v = \sqrt{Rg\tan\theta} = \sqrt{(5.50 \text{ m})(9.80 \text{ m/s}^2)\tan 30.0°} = 5.58 \text{ m/s}.$$

Then the period is $T = \dfrac{2\pi R}{v} = \dfrac{2\pi(5.50 \text{ m})}{5.58 \text{ m/s}} = 6.19 \text{ s}.$

(c) The net force is proportional to m so in $\Sigma\vec{F} = m\vec{a}$ the mass divides out and the angle for a given rate of rotation is independent of the mass of the passengers.

Reflect: The person moves in a horizontal circle so the acceleration is horizontal. The net inward force required for circular motion is produced by a component of the force exerted on the seat by the rod.

6.9. Set Up: At the safe speed no friction force is required. Consider a car of mass m. The free-body diagram for the car is given in Figure 6.9. Use coordinates where $+x$ is toward the center of the horizontal turn. Each turn is one-quarter of a circle so the radius R is given by $0.25 \text{ mi} = \frac{1}{4}(2\pi R)$. $9°12'$ is $9.2°$.

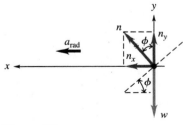

Figure 6.9

Solve: $R = 4(0.25 \text{ mi})/2\pi = 0.1592 \text{ mi} = 256 \text{ m}$

$\Sigma F_y = ma_y$ gives $n_y = w$ so $n = \dfrac{mg}{\cos\phi}$

$$\Sigma F_x = ma_x \text{ gives } n_x = ma_{\text{rad}} \text{ so } n\sin\phi = m\frac{v^2}{R}$$

$$\left(\frac{mg}{\cos\phi}\right)\sin\phi = m\frac{v^2}{R} \text{ and } v = \sqrt{Rg\tan\phi} = \sqrt{(256\text{ m})(9.80\text{ m/s}^2)\tan9.2°} = 20.0\text{ m/s} = 44.7\text{ mph}$$

Reflect: Race cars travel *much* faster than this. The track must exert a large friction force on the cars to maintain their circular motion.

6.11. Set Up: The period is $T = 60.0$ s and $T = \dfrac{2\pi R}{v}$. The apparent weight of a person is the normal force exerted on him by the seat he is sitting on. His acceleration is $a_{\text{rad}} = v^2/R$, directed toward the center of the circle. The passenger has mass $m = w/g = 90.0$ kg.

Solve: (a) $v = \dfrac{2\pi R}{T} = \dfrac{2\pi(50\text{ m})}{60.0\text{ s}} = 5.24$ m/s. Note that $a_{\text{rad}} = \dfrac{v^2}{R} = \dfrac{(5.24\text{ m/s})^2}{50.0\text{ m}} = 0.549$ m/s^2.

(b) The free-body diagram for the person at the top of his path is given in Figure 6.11a. The acceleration is downward, so take $+y$ downward. $\Sigma F_y = ma_y$ gives $mg - n = ma_{\text{rad}}$.

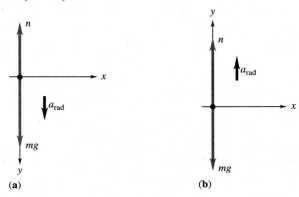

Figure 6.11

$$n = m(g - a_{\text{rad}}) = (90.0\text{ kg})(9.80\text{ m/s}^2 - 0.549\text{ m/s}^2) = 833\text{ N}.$$

The free-body diagram for the person at the bottom of his path is given in Figure 6.11b. The acceleration is upward, so take $+y$ upward. $\Sigma F_y = ma_y$ gives $n - mg = ma_{\text{rad}}$ and $n = m(g + a_{\text{rad}}) = 931$ N.

(c) Apparent weight $= 0$ means $n = 0$ and $mg = ma_{\text{rad}}$.

$$g = \frac{v^2}{R} \text{ and } v = \sqrt{gR} = 22.1\text{ m/s}.$$

The time for one revolution would be

$$T = \frac{2\pi R}{v} = \frac{2\pi(50.0\text{ m})}{22.1\text{ m/s}} = 14.2\text{ s}.$$

Note that $a_{\text{rad}} = g$.

(d) $n = m(g + a_{\text{rad}}) = 2mg = 2(882\text{ N}) = 1760$ N, twice his true weight.

Reflect: At the top of his path his apparent weight is less than his true weight and at the bottom of his path his apparent weight is greater than his true weight.

6.13. Set Up: $R = 0.700$ m. A 45° angle is $\frac{1}{8}$ of a full rotation, so in $\frac{1}{2}$ s a hand travels through a distance of $\frac{1}{8}(2\pi R)$. In (c) use coordinates where $+y$ is upward, in the direction of $\vec{a}_{\text{rad}}$ at the bottom of the swing.

Solve: (a) $v = \frac{1}{8}\left(\frac{2\pi R}{0.50 \text{ s}}\right) = 1.10 \text{ m/s}$ and $a_{\text{rad}} = \frac{v^2}{R} = \frac{(1.10 \text{ m/s})^2}{0.700 \text{ m}} = 1.73 \text{ m/s}^2$

(b) The free-body diagram is shown in Figure 6.13. F is the force exerted by the blood vessel.

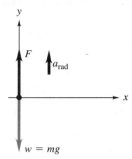

Figure 6.13

(c) $\sum F_y = ma_y$ gives $F - w = ma_{\text{rad}}$ and

$$F = m(g + a_{\text{rad}}) = (1.00 \times 10^{-3} \text{ kg})(9.80 \text{ m/s}^2 + 1.73 \text{ m/s}^2) = 1.15 \times 10^{-2} \text{ N, upward}$$

(d) When the arm hangs vertically and at rest, $a_{\text{rad}} = 0$ so $F = w = mg = 9.8 \times 10^{-3} \text{ N}$.

Reflect: The acceleration of the hand is only about 20% of g, so the increase in the force on the blood drop when the arm swings is about 20%.

6.15. Set Up: The discussion at the end of Section 6.2 relates walking speed to the length R of the leg and g, the acceleration due to gravity on the moon.

Solve: (a) $v_{\text{max}} = \sqrt{gR} = \sqrt{(1.67 \text{ m/s}^2)(0.95 \text{ m})} = 1.3 \text{ m/s}$

(b) On earth $g = 9.80 \text{ m/s}^2$ so $v_{\text{max}} = \sqrt{(9.80 \text{ m/s}^2)(0.95 \text{ m})} = 3.1 \text{ m/s}$

Reflect: He can walk faster on earth because the gravity force is greater there.

6.19. Set Up: The gravitational force between two objects is $F_g = G\frac{m_1 m_2}{r^2}$. The mass of the earth is $m_E = 5.97 \times 10^{24} \text{ kg}$ and the mass of the sun is $m_S = 1.99 \times 10^{30} \text{ kg}$. The distance from the earth to the sun is $r_{SE} = 1.50 \times 10^{11} \text{ m}$ and the distance from the earth to the moon is $r_{EM} = 3.84 \times 10^8 \text{ m}$.

Solve: $F_{SM} = G\frac{m_S m_M}{r_{SM}^2} = G\frac{m_S m_M}{r_{SE}^2}$. $F_{EM} = G\frac{m_E m_M}{r_{EM}^2}$.

$$\frac{F_{SM}}{F_{EM}} = \left(\frac{m_S}{m_E}\right)\left(\frac{r_{EM}}{r_{SE}}\right)^2 = \left(\frac{1.99 \times 10^{30} \text{ kg}}{5.97 \times 10^{24} \text{ kg}}\right)\left(\frac{3.84 \times 10^8 \text{ m}}{1.50 \times 10^{11} \text{ m}}\right)^2 = 2.18$$

It is more accurate to say the moon orbits the sun.

Reflect: The sun is farther away but has a much greater mass than the earth.

6.21. Set Up: When we take the ratio of the forces F_A and F_B at $r_A = 10 \times 10^9 \text{ km} = 1.0 \times 10^{13} \text{ m}$ and $r_B = 130 \times 10^9 \text{ km} = 1.3 \times 10^{14} \text{ m}$, the masses of the two objects divides out.

Solve: (a) The orbit is sketched in Figure 6.21. $r_B/r_A = 13$

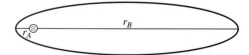

Figure 6.21

(b) $F_A = G\frac{m_1 m_2}{r_A^2}$, $F_B = G\frac{m_1 m_2}{r_B^2}$ so $\frac{F_A}{F_B} = \left(\frac{r_B}{r_A}\right)^2 = (13)^2 = 169$

Reflect: The force is greater when Sedna is closer to the sun.

6.23. Set Up: The two masses are sketched in Figure 6.23a. Let the $3M$ mass be to the right of M. For the net force to be zero, the forces F_1 and F_2 due to M and $3M$ must be equal in magnitude and opposite in direction. The two forces will be opposite in direction only at points between the two masses. Consider a point P that is a distance d from M, so 1.00 m $- d$ from $3M$. Let the third object, at P, have mass m.

Figure 6.23

Solve: The three masses and the forces on m are shown in Figure 6.23b. $F_1 = F_2$ gives $G\dfrac{mM}{d^2} = G\dfrac{m(3M)}{(1.00\text{ m} - d)^2}$

$3d^2 = (1.00\text{ m} - d)^2$ so $\sqrt{3}\,d = 1.00$ m $- d$ and $d = \dfrac{1.00\text{ m}}{1 + \sqrt{3}} = 0.366$ m $= 36.6$ cm

Reflect: Point P is closer to the smaller mass.

6.27. Set Up: The circumference c is related to the radius R_p of the planet by $c = 2\pi R_p$. Take $+y$ downward and use the measured motion to find $a_y = g$, the acceleration due to gravity at the surface of the planet. $m_E = 5.97 \times 10^{24}$ kg

Solve: **(a)** $v_{0y} = 0$, $t = 0.811$ s, $y - y_0 = 5.00$ m

$y = y_0 + v_{0y}t + \frac{1}{2}a_y t^2$ gives $a_y = \dfrac{2(y - y_0)}{t^2} = \dfrac{2(5.00\text{ m})}{(0.811\text{ s})^2} = 15.2$ m/s^2

$$R_p = \frac{c}{2\pi} = \frac{6.24 \times 10^7 \text{ m}}{2\pi} = 9.93 \times 10^6 \text{ m}$$

$$g = \frac{Gm_p}{R_p^2} \text{ so } m_p = \frac{gR_p^2}{G} = \frac{(15.2\text{ m/s}^2)(9.93 \times 10^6 \text{ m})^2}{6.673 \times 10^{-11}\text{ N} \cdot \text{m}^2/\text{kg}^2} = 2.25 \times 10^{25} \text{ kg}$$

(b) $m_p = 3.76 m_E$

Reflect: $R_p = 1.56 R_E$ and $m_p = 3.76 m_E$ so $g_p = \dfrac{3.76}{(1.56)^2} g_E = 15.2$ m/s^2, which checks.

6.29. Set Up: The mass of the probe on earth is $m = \dfrac{w_E}{g_E} = \dfrac{3120\text{ N}}{9.80\text{ m/s}^2} = 318.4$ kg. The radius of Titan is $R_T = \frac{1}{2}(5150\text{ km}) = 2.575 \times 10^6$ m.

Solve: $w = G\dfrac{mm_T}{R_T^2} = \dfrac{(6.673 \times 10^{-11}\text{ N} \cdot \text{m}^2/\text{kg}^2)(318.4\text{ kg})(1.35 \times 10^{23}\text{ kg})}{(2.575 \times 10^6 \text{ m})^2} = 433$ N

Reflect: The mass of the probe is independent of its location.

6.37. Set Up: The rotational period of the earth is 1 day $= 8.64 \times 10^4$ s. The radius of the satellite's path is $r = h + R_E$, where h is its height above the surface of the earth and $R_E = 6.38 \times 10^6$ m. v and r are related by applying $\Sigma\vec{F} = m\vec{a}$ to the motion of the satellite. $m_E = 5.97 \times 10^{24}$ kg

Solve: **(a)** 24 h $= 8.64 \times 10^4$ s

(b)-(c) The free-body diagram for the satellite is given in Figure 6.37a. $F_g = ma_{rad}$ gives $G\dfrac{m_E m}{r^2} = m\dfrac{v^2}{r}$ and $v^2 = \dfrac{Gm_E}{r}$. $T = \dfrac{2\pi r}{v}$ so $v = \dfrac{2\pi r}{T}$ and $\dfrac{4\pi^2 r^2}{T^2} = \dfrac{Gm_E}{r}$.

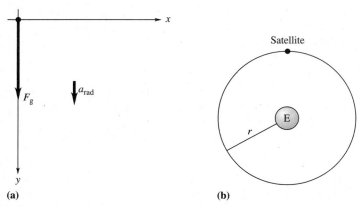

Figure 6.37

$$r = \left(\frac{Gm_E T^2}{4\pi^2}\right)^{1/3} = \left[\frac{(6.673 \times 10^{-11}\ \text{N}\cdot\text{m}^2/\text{kg}^2)(5.97 \times 10^{24}\ \text{kg})(8.64 \times 10^4\ \text{s})^2}{4\pi^2}\right]^{1/3} = 4.22 \times 10^7\ \text{m}$$

$$h = r - R_E = 3.58 \times 10^7\ \text{m} = 3.58 \times 10^4\ \text{km}$$

(d) $\dfrac{r}{R_E} = 5.6$. The sketch of the earth and orbit of the satellite is given in Figure 6.37b.

Reflect: The orbital period is proportional to $r^{3/2}$. To achieve a period as large as 24 h, the satellite is at a large altitude above the earth's surface.

6.41. Set Up: The mass of the earth is $m_E = 5.97 \times 10^{24}$ kg.
Solve: (a) The radius of the event horizon for a black hole of mass m_E is

$$R_S = \frac{2Gm_E}{c^2} = \frac{2(6.673 \times 10^{-11}\ \text{N}\cdot\text{m}^2/\text{kg}^2)(5.97 \times 10^{24}\ \text{kg})}{(3.00 \times 10^8\ \text{m/s})^2} = 8.85 \times 10^{-3}\ \text{m}$$

(b) Your mass is $m = w/g = 76.5$ kg.

$$F_g = G\frac{mm_E}{R_S^2} = \frac{(6.673 \times 10^{-11}\ \text{N}\cdot\text{m}^2/\text{kg}^2)(76.5\ \text{kg})(5.97 \times 10^{24}\ \text{kg})}{(8.85 \times 10^{-3}\ \text{m})^2} = 3.89 \times 10^{20}\ \text{N}$$

Reflect: The calculation in (b) assumes that you can be treated as a point mass, with all of your mass at the event horizon.

6.47. Set Up: Your friend slides toward you when the friction force on her isn't sufficient to make her turn as you turn. Set $f_s = \mu_s n$ and apply $\Sigma\vec{F} = m\vec{a}$ to your friend. Set her acceleration equal to that of the car, $a_{rad} = \dfrac{v^2}{r}$.

Solve: **(a)** In the absence of friction your friend moves in a straight line, so turn to the right, toward her.
(b) The free-body diagram for your friend is given in Figure 6.47.

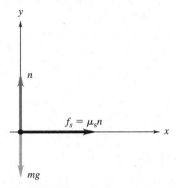

Figure 6.47

$\sum F_y = ma_y$ gives $n = mg$. $\sum F_x = ma_x$ gives $\mu_s n = ma_{rad}$. $\mu_s mg = m\dfrac{v^2}{R}$ and

$$R = \frac{v^2}{\mu_s g} = \frac{(15 \text{ m/s})^2}{(0.55)(9.80 \text{ m/s}^2)} = 42 \text{ m}.$$

Reflect: The larger the coefficient of friction, the smaller must be the radius of the turn.

6.49. Set Up: The person moves in a horizontal circle of radius $r = 2.5$ m. Set the static friction force equal to its maximum value, $f_s = \mu_s n$. The person has an acceleration

$$a_{rad} = \frac{v^2}{r},$$

directed toward the center of the circle. The period is

$$T = \frac{1}{0.60 \text{ rev/s}} = 1.67 \text{ s}$$

and the person has speed

$$v = \frac{2\pi r}{T} = 9.41 \text{ m/s}.$$

Solve: **(a)** The free-body diagram is given in Figure 6.49. The diagram is for when the wall is to the right of her, so the center of the cylinder is to the left.

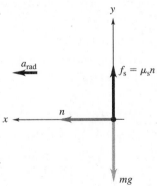

Figure 6.49

(b) $\Sigma F_x = ma_x$ gives $n = m\dfrac{v^2}{r}$. $\Sigma F_y = ma_y$ gives $f_s = mg$, so $\mu_s n = mg$. Combining these two equations gives

$$\mu_s m \frac{v^2}{r} = mg \text{ and}$$

$$\mu_s = \frac{gr}{v^2} = \frac{(9.80 \text{ m/s}^2)(2.5 \text{ m})}{(9.41 \text{ m/s})^2} = 0.28.$$

(c) The mass m of the person divides out of the equation for μ_s; the answer to (b) does not depend on the mass of the person.

Reflect: The greater the rotation rate the larger the normal force exerted by the wall and the larger the friction force. Therefore, for smaller μ_s the rotation rate must be larger.

6.53. Set Up: Use the measurements of the motion of the rock to calculate g_M, the value of g on Mongo. Take $+y$ upward. When the stone returns to the ground its velocity is 12.0 m/s, downward. $g_M = G\dfrac{m_M}{R_M^2}$. The radius of Mongo is

$$R_M = \frac{c}{2\pi} = \frac{2.00 \times 10^8 \text{ m}}{2\pi} = 3.18 \times 10^7 \text{ m}.$$

The ship moves in an orbit of radius $r = 3.18 \times 10^7 \text{ m} + 3.00 \times 10^7 \text{ m} = 6.18 \times 10^7 \text{ m}$. For the ship, $F_g = ma_{rad}$ and $T = \dfrac{2\pi r}{v}$.

Solve: **(a)** $v_{0y} = +12.0$ m/s, $v_y = -12.0$ m/s, $a_y = -g_M$ and $t = 8.00$ s. $v_y = v_{0y} + a_y t$ gives

$$-g_M = \frac{v_y - v_{0y}}{t} = \frac{-12.0 \text{ m/s} - 12.0 \text{ m/s}}{8.00 \text{ s}} \text{ and } g_M = 3.00 \text{ m/s}^2.$$

$$m_M = \frac{g_M R_M^2}{G} = \frac{(3.00 \text{ m/s}^2)(3.18 \times 10^7 \text{ m})^2}{6.673 \times 10^{-11} \text{ N} \cdot \text{m}^2/\text{kg}^2} = 4.55 \times 10^{25} \text{ kg}$$

(b) $F_g = ma_{rad}$ gives $G\dfrac{m_M m}{r^2} = m\dfrac{v^2}{r}$ and $v^2 = \dfrac{Gm_M}{r}$.

$$T = \frac{2\pi r}{v} = 2\pi r\sqrt{\frac{r}{Gm_M}} = \frac{2\pi r^{3/2}}{\sqrt{Gm_M}} = \frac{2\pi(6.18 \times 10^7 \text{ m})^{3/2}}{\sqrt{(6.673 \times 10^{-11} \text{ N} \cdot \text{m}^2/\text{kg}^2)(4.55 \times 10^{25} \text{ kg})}}$$

$$T = 5.54 \times 10^4 \text{ s} = 15.4 \text{ h}$$

Reflect: $R_M = 5.0R_E$ and $m_M = 7.6m_E$, so $g_M = \dfrac{7.6}{(5.0)^2}g_E = 0.30g_E$, which agrees with the value calculated in part (a).

6.55. Set Up: The volume of a sphere is $V = \frac{4}{3}\pi R^3$. $R_S = 8.5 \times 10^5$ m. The mass m_S of Sedna is $m_S = \rho_S V_S$, where $\rho_S = 2.1 \times 10^3$ kg/m^3. The mass of the astronaut is

$$m = \frac{w}{g} = \frac{675 \text{ N}}{9.80 \text{ m/s}^2} = 68.9 \text{ kg}$$

Solve: $m_S = \rho_S V_S = \rho_S \frac{4}{3}\pi R^3 = (2.1 \times 10^3 \text{ kg/m}^3)(\frac{4}{3}\pi)(8.5 \times 10^5 \text{ m})^3 = 5.40 \times 10^{21} \text{ kg}$

$$g = G\frac{m_S}{R_S^2} = \frac{(6.673 \times 10^{-11} \text{ N} \cdot \text{m}^2/\text{kg}^2)(5.40 \times 10^{21} \text{ kg})}{(8.5 \times 10^5 \text{ m})^2} = 0.499 \text{ m/s}^2$$

Then $w = mg = (68.9 \text{ kg})(0.499 \text{ m/s}^2) = 34.4$ N.

Reflect: g is much less on Sedna so w is much less there.

6.57. Set Up: The block moves in a horizontal circle of radius $r = \sqrt{(1.25 \text{ m})^2 - (1.00 \text{ m})^2} = 0.75$ m. Each string makes an angle θ with the vertical.

$$\cos\theta = \frac{1.00 \text{ m}}{1.25 \text{ m}}, \text{ so } \theta = 36.9°.$$

The block has acceleration $a_{rad} = v^2/r$, directed to the left in the figure in the problem.

Solve: (a) The free-body diagram for the block is given in Figure 6.57. $\Sigma F_y = ma_y$ gives $T_u\cos\theta - T_l\cos\theta - mg = 0$.

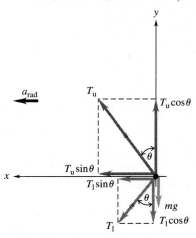

Figure 6.57

$$T_l = T_u - \frac{mg}{\cos\theta} = 80.0 \text{ N} - \frac{(4.00 \text{ kg})(9.80 \text{ m/s}^2)}{\cos 36.9°} = 31.0 \text{ N}$$

(b) $\Sigma F_x = ma_x$ gives $(T_u + T_l)\sin\theta = m\dfrac{v^2}{r}$.

$$v = \sqrt{\frac{r(T_u + T_l)\sin\theta}{m}} = \sqrt{\frac{(0.75 \text{ m})(80.0 \text{ N} + 31.0 \text{ N})\sin 36.9°}{4.00 \text{ kg}}} = 3.53 \text{ m/s}$$

Reflect: The tension in the upper string must be greater than the tension in the lower string so that together they produce an upward component of force that balances the weight of the block.

6.59. Set Up: The planet has mass $14m_E = 8.36 \times 10^{25}$ kg and density $\rho = 5.5 \times 10^3$ kg/m^3. The volume of a sphere is $V = \frac{4}{3}\pi R^3$.

Solve: (a) $\rho = \dfrac{m}{V} = \dfrac{m}{\frac{4}{3}\pi R^3}$ so

$$R = \left(\frac{3m}{4\pi\rho}\right)^{1/3} = \left[\frac{3(8.36 \times 10^{25} \text{ kg})}{4\pi(5.5 \times 10^3 \text{ kg/m}^3)}\right]^{1/3} = 1.54 \times 10^7 \text{ m}$$

(b) $g = G\dfrac{m}{R^2} = \dfrac{(6.673 \times 10^{-11} \text{ N} \cdot \text{m}^2/\text{kg}^2)(8.36 \times 10^{25} \text{ kg})}{(1.54 \times 10^7 \text{ m})^2} = 23.5 \text{ m/s}^2 = 2.4g$

Reflect: $g = Gm\left(\dfrac{4\pi\rho}{3m}\right)^{2/3} = G\left(\dfrac{4\pi\rho}{3}\right)^{2/3}m^{1/3}$, so g is proportional to $m^{1/3}$ for planets of the same density. m is larger by a factor of 14 compared to earth, so g is larger by a factor of $(14)^{1/3}$.

6.61. Set Up: The earth has mass $m_E = 5.97 \times 10^{24}$ kg, radius $R_E = 6.38 \times 10^6$ m and rotational period $T = 24$ hr $= 8.64 \times 10^4$ s. Use coordinates for which the $+y$ direction is toward the center of the earth. The free-body diagram for Sneezy at the equator is given in Figure 6.61.

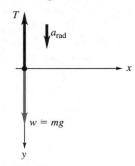

Figure 6.61

Solve: At the north pole Sneezy has $a = 0$ and $T = w = 475.0$ N, the gravitational force exerted by the earth. Sneezy has mass $w/g = 48.47$ kg.

At the equator Sneezy is traveling in a circular path and has acceleration

$$a_{rad} = \frac{4\pi^2 R}{T^2} = \frac{4\pi^2(6.38 \times 10^6 \text{ m})}{(8.64 \times 10^4 \text{ s})^2} = 0.0337 \text{ m/s}^2$$

$\sum F_y = ma_y$ gives $w - T = ma_{rad}$ and

$$T = w - ma_{rad} = m(g - a_{rad}) = (48.47 \text{ kg})(9.80 \text{ m/s}^2 - 0.0337 \text{ m/s}^2) = 473.4 \text{ N}$$

Reflect: At the equator the object has an inward acceleration and the outward tension is less than the true weight, since there is a net inward force.

WORK AND ENERGY

Problems 3, 7, 11, 13, 19, 21, 23, 27, 29, 31, 35, 41, 45, 47, 49, 53, 57, 59, 61, 63, 67, 73, 75, 77, 79, 81, 83, 85, 87, 93, 97, 99, 101, 105, 111

7.3. Set Up: Use $W = F_{\parallel}s = (F\cos\phi)s$ with $\phi = 15.0°$.
Solve: $W = (F\cos\phi)s = (180\text{ N})(\cos 15.0°)(300.0\text{ m}) = 5.22 \times 10^4\text{ J}$
Reflect: Since $\cos 15.0° \approx 0.97$, the relatively small angle of 15.0° allows the boat to apply approximately 97% of the 180 N force to pulling the skier.

7.7. Set Up: In order to move the crate at constant velocity, the worker must apply a force that equals the force of friction, $F_{\text{worker}} = f_k = \mu_k n$. Each force can be used in the relation $W = F_{\parallel}s = (F\cos\phi)s$ for parts (b) through (d). For part (e), apply the net work relation as $W_{\text{net}} = W_{\text{worker}} + W_{\text{grav}} + W_n + W_f$.
Solve: (a) The magnitude of the force the worker must apply is:

$$F_{\text{worker}} = f_k = \mu_k n = \mu_k mg = (0.25)(30.0\text{ kg})(9.80\text{ m/s}^2) = 74\text{ N}$$

(b) Since the force applied by the worker is horizontal and in the direction of the displacement, $\phi = 0°$ and the work is:

$$W_{\text{worker}} = (F_{\text{worker}}\cos\phi)s = [(74\text{ N})(\cos 0°)](4.5\text{ m}) = +333\text{ J}$$

(c) Friction acts in the direction opposite of motion, thus $\phi = 180°$ and the work of friction is:

$$W_f = (f_k\cos\phi)s = [(74\text{ N})(\cos 180°)](4.5\text{ m}) = -333\text{ J}$$

(d) Both gravity and the normal force act perpendicular to the direction of displacement. Thus, neither force does any work on the crate and $W_{\text{grav}} = W_n = 0.0\text{ J}$.
(e) Substituting into the net work relation, the net work done on the crate is:

$$W_{\text{net}} = W_{\text{worker}} + W_{\text{grav}} + W_n + W_f = +333\text{ J} + 0.0\text{ J} + 0.0\text{ J} - 333\text{ J} = 0.0\text{ J}$$

Reflect: The net work done on the crate is zero because the two contributing forces, F_{worker} and F_f, are equal in magnitude and opposite in direction.

7.11. Set Up: Since the speed is constant, the acceleration and the net force on the monitor are zero. Use this fact to develop expressions for the friction force, f_k, and the normal force, n. Then use $W = F_{\parallel}s = (F\cos\phi)s$ to calculate W.

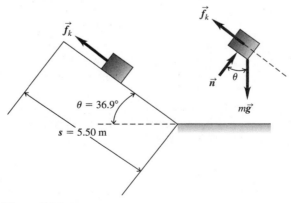

Figure 7.11

Solve: (a) Summing forces along the incline,

$$\sum F = ma = 0 = f_k - mg\sin\theta$$

giving $f_k = mg\cos\theta$, directed up the incline. Substituting,

$$W_f = (f_k\cos\phi)s = [(mg\sin\theta)\cos\phi]s$$
$$= [(10.0\text{ kg})(9.80\text{ m/s}^2)(\sin 36.9°)](\cos 0°)(5.50\text{ m})$$
$$= +324\text{ J}$$

(b) The gravity force is downward and the displacement is directed up the incline so $\phi = 126.9°$.

$$W_{grav} = (10.0\text{ kg})(9.80\text{ m/s}^2)(\cos 126.9°)(5.50\text{ m}) = -324\text{ J}$$

(c) The normal force, n, is perpendicular to the displacement and thus does zero work.
Reflect: Friction does positive work and gravity does negative work. The net work done is zero.

7.13. Set Up: Use $K = \frac{1}{2}mv^2$ and solve for m.
Solve: $m = 2K/v^2 = 2(1960\text{ J})/(965\text{ m/s})^2 = 4.21 \times 10^{-3}\text{ kg} = 4.21\text{ g}$
Reflect: The kinetic energy of an object is proportional to the mass of the object.

7.19. Set Up: Use the work-kinetic energy theorem: $W_{net} = K_f - K_i = F_{net}s$. Since the net force is due to friction, $F_{net}s = -f_k s = -\mu_k mgs$. Also, since the car stops, $K_f = 0$.
Solve: (a) $W_{net} = K_f - K_i = F_{net}s$ gives $-\frac{1}{2}mv_i^2 = -\mu_k mgs$. Solving for the distance,

$$s = \frac{v_i^2}{2\mu_k g} = \frac{(23.0\text{ m/s})^2}{2(0.700)(9.80\text{ m/s}^2)} = 38.6\text{ m}$$

(b) Since s is proportional to v_i^2, doubling v_i increases s by a factor of 4; s therefore becomes 154 m.
(c) The original kinetic energy was converted into thermal energy by the negative work of friction.
Reflect: To stop the car friction must do negative work equal in magnitude to the initial kinetic energy of the car.

7.21. Set Up: From the work-energy relation, $W = W_{grav} = \Delta K_{rock}$ or $F_{\parallel}s = K_f - K_i$. As the rock rises, the gravitational force, $F = mg$, does work on the rock. Since this force acts in the direction opposite to the motion and displacement, s, the work is negative.
Solve: (a) Applying $F_{\parallel}s = K_f - K_i$ we obtain:

$$-mgh = \frac{1}{2}mv_f^2 - \frac{1}{2}mv_i^2$$

Dividing by m and solving for v_i, $v_i = \sqrt{v_f^2 + 2gh}$. Substituting $h = 15.0$ m and $v_f = 25.0$ m/s,

$$v_i = \sqrt{(25.0\text{ m/s})^2 + 2(9.80\text{ m/s}^2)(15.0\text{ m})} = 30.3\text{ m/s}$$

(b) Solve the same work-energy relation for h. At the maximum height $v_f = 0$.

$$-mgh = \frac{1}{2}mv_f^2 - \frac{1}{2}mv_i^2$$
$$h = \frac{v_i^2 - v_f^2}{2g} = \frac{(30.3\text{ m/s})^2 - (0.0\text{ m/s})^2}{2(9.80\text{ m/s}^2)} = 46.8\text{ m}$$

Reflect: Note that the weight of 20 N was never used in the calculations because both gravitational potential and kinetic energy are proportional to mass, m. Thus any object, that attains 25.0 m/s at a height of 15.0 m, must have an initial velocity of 30.3 m/s. As the rock moves upward gravity does negative work and this reduces the kinetic energy of the rock.

7.23. Set Up: Use $W_{net} = W_f = -f_k s = -\mu_k mgs$ and $W_{net} = K_f - K_i$. The skier stops, so $K_f = 0$.
Solve: (a) Setting the two expression for net work equal, $W_{net} = -\mu_k mgs = -\frac{1}{2}mv_i^2$. Solving for the coefficient of kinetic friction,

$$\mu_k = \frac{v_i^2}{2gs} = \frac{(12.0\text{ m/s})^2}{2(9.80\text{ m/s}^2)(184\text{ m})} = 3.99 \times 10^{-2}$$

(b) The mass m of the skier divides out and μ_k is independent of m. If v_i is doubled while s is constant then μ_k increases by a factor of 4; $\mu_k = 0.160$.

Reflect: To stop the skier friction does negative work that is equal in magnitude to the initial kinetic energy of the skier.

7.27. Set Up: Solve $F_{\text{on spring}} = kx$ for x to determine the length of stretch and use $W_{\text{on spring}} = +\frac{1}{2}kx^2$ to assess the corresponding work.

Solve: $x = \dfrac{F_{\text{on spring}}}{k} = \dfrac{15.0 \text{ N}}{300.0 \text{ N/m}} = 0.0500 \text{ m}$

The new length will be $0.240 \text{ m} + 0.0500 \text{ m} = 0.290 \text{ m}$. The corresponding work done is,

$$W_{\text{on spring}} = \frac{1}{2}(300.0 \text{ N/m})(0.0500 \text{ m})^2 = 0.375 \text{ J}$$

Reflect: $F_{\text{on spring}}$ is always the force applied to one end of the spring, thus we did not need to double the 15.0 N force. Consider a free-body diagram of a spring at rest; forces of equal magnitude and opposite direction are always applied to both ends of every section of the spring examined.

7.29. Set Up: The work done is the area under the graph of $F_{\text{on spring}}$ versus x between the initial and final positions.
Solve: (a) The area between $x = 0$ and $x = 5.0$ cm is a right triangle. The work is thus calculated as:

$$W_{\text{on spring}} = \frac{1}{2}(5.0 \times 10^{-2} \text{ m})(250 \text{ N}) = 6.2 \text{ J}$$

(b) Since the area between $x = 2.0$ cm and $x = 7.0$ cm is a trapezoid, the work is calculated as:

$$W_{\text{on spring}} = \frac{1}{2}(100 \text{ N} + 350 \text{ N})(0.070 \text{ m} - 0.020 \text{ m}) = 11.2 \text{ J}$$

Reflect: The force is larger for $x = 7.0$ cm than for 5.0 cm and the work done in (b) is greater than that done in (a), even though the displacement of the end of the spring is 5.0 cm in each case.

7.31. Set Up: Use $\Delta U_{\text{grav}} = mg(y_f - y_i)$.
Solve: $\Delta U_{\text{grav}} = (72 \text{ kg})(9.80 \text{ m/s}^2)(0.60 \text{ m}) = 420 \text{ J}$.
Reflect: This gravitational potential energy comes from elastic potential stored in his tensed muscles.

7.35. Set Up: Use $F_{\text{on tendon}} = kx$. In part (a), $F_{\text{on tendon}}$ equals mg, the weight of the object suspended from it. In part (b), also apply $U_{\text{el}} = \frac{1}{2}kx^2$ to calculate the stored energy.
Solve: (a) $k = \dfrac{F_{\text{on tendon}}}{x} = \dfrac{(0.250 \text{ kg})(9.80 \text{ m/s}^2)}{0.0123 \text{ m}} = 199 \text{ N/m}$

(b) $x = \dfrac{F_{\text{on tendon}}}{k} = \dfrac{138 \text{ N}}{199 \text{ N/m}} = 0.693 \text{ m} = 69.3 \text{ cm}; \; U_{\text{el}} = \frac{1}{2}(199 \text{ N/m})(0.693 \text{ m})^2 = 47.8 \text{ J}$

Reflect: The 250 g object has a weight of 2.45 N. The 138 N force is much larger than this and stretches the tendon a much greater distance.

7.41. Set Up: Set 20% of the food energy (in joules) equal to the total work you do in jumping upward a distance of $h = 50$ cm, N times.
Solve:

$$0.20(500 \text{ food cal})\left(\frac{4186 \text{ J}}{1 \text{ food cal}}\right) = N(mgh)$$

$$N = \frac{(0.20)(500)(4186 \text{ J})}{(75 \text{ kg})(9.80 \text{ m/s}^2)(0.50 \text{ m})} = 1.1 \times 10^3 \text{ jumps}$$

The total time it would take is thus $N(2.0 \text{ s}) = 2.2 \times 10^3 \text{ s} = 37$ minutes.
Reflect: This would be a very strenuous workout even for a very fit person.

7.45. Set Up: Let θ be the angle of the initial speed above the horizontal. In parts (a)-(c), θ has values of $90°$, $-90°$, and $0°$, respectively. Use $K_f + U_f = K_i + U_i$ with $K_f = \frac{1}{2}(w/g)v_f^2$ and $K_i = \frac{1}{2}(w/g)v_0^2$. Let $y_i = h$ and $y_f = 0$; this gives $\frac{1}{2}(w/g)v_f^2 = \frac{1}{2}(w/g)v_0^2 + wh$.

Solve: (a)-(d) $v_f = \sqrt{v_0^2 + 2gh}$. **(e)** Since K and U are both proportional to the weight $w = mg$, w divides out of the expression and v_f is unaffected by a change in weight.

Reflect: The initial kinetic energy depends only on the initial speed and is independent of the direction of the initial velocity.

7.47. Set Up: Use $K_f + U_f = K_i + U_i$. Let $y_i = 0$ and $y_f = h$ and note that $U_i = 0$ while $K_f = 0$ at the maximum height. Consequently, conservation of energy becomes $mgh = \frac{1}{2}mv_i^2$.

Solve: (a) $v_i = \sqrt{2gh} = \sqrt{2(9.80 \text{ m/s}^2)(0.20 \text{ m})} = 2.0$ m/s.

(b) $K_i = mgh = (0.50 \times 10^{-6} \text{ kg})(9.80 \text{ m/s}^2)(0.20 \text{ m}) = 9.8 \times 10^{-7}$ J

$$\frac{K_i}{m} = \frac{9.8 \times 10^{-7} \text{ J}}{0.50 \times 10^{-6} \text{ kg}} = 2.0 \text{ J/kg}$$

(c) The human can jump to a height of

$$h_h = h_f\left(\frac{l_h}{l_f}\right) = (0.20 \text{ m})\left(\frac{2.0 \text{ m}}{2.0 \times 10^{-3} \text{ m}}\right) = 200 \text{ m}.$$

To attain this height, he would require a takeoff speed of: $v_i = \sqrt{2gh} = \sqrt{2(9.80 \text{ m/s}^2)(200 \text{ m})} = 63$ m/s.

(d) The human's kinetic energy per kilogram,

$$\frac{K_i}{m} = gh = (9.80 \text{ m/s}^2)(0.60 \text{ m}) = 5.9 \text{ J/kg}.$$

(e) Reflect: The flea stores the energy in its tensed legs.

7.49. Set Up: Apply $K_f + U_f = K_i + U_i$ to the motion of the stone on the moon with $U_i = 0$ and $K_f = 0$ at the maximum height. The conservation equation becomes $U_f = K_i$ with $v_i = v_{moon,i} = v$ and $K_i = K_{moon,i} = \frac{1}{2}mv^2$. Setting $y_f = h$ and $y_i = 0$ gives $\frac{1}{2}mv^2 = mgh$. Note that v^2/g is the same on the earth and moon and let subscript 1 refer to the moon and subscript 2 refer to the earth.

Solve: $v_1^2/g_1 = v_2^2/g_2$; $v_2 = v_1\sqrt{\dfrac{g_2}{g_1}} = v\sqrt{\dfrac{9.80 \text{ m/s}^2}{1.67 \text{ m/s}^2}} = 2.42v$

Reflect: Gravity is stronger on earth so a larger initial speed is needed in order to achieve the same height as on the moon.

7.53. Set Up: Use $K_f + U_f = K_i + U_i$ with $K_i = K_f = 0$. If we let $y_f = 0$ then $y_i = x$, the amount the spring will be compressed. This results in $\frac{1}{2}kx^2 = mgx$.

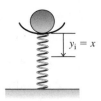

$y_i = x$

Figure 7.53

Solve: $x = \dfrac{2mg}{k} = \dfrac{2(12.0 \text{ N})}{325 \text{ N/m}} = 7.38 \times 10^{-2}$ m $= 7.38$ cm

Reflect: At the maximum compression the magnitude of the spring force is twice the weight of the object and the system is not in equilibrium.

7.57. Set Up: Use $E_{mech} = K + U$ with $y_f = 0$, $y_i = h$, and $K_i = U_f = 0$. Conservation of energy thus becomes: $E_{mech,i} - E_{mech,f} = mgy_i - \frac{1}{2}mv_f^2$.

Solve: $E_{mech,i} - E_{mech,f} = (12.0 \times 10^{-3} \text{ kg})[(9.80 \text{ m/s}^2)(2.50 \text{ m}) - \frac{1}{2}(3.20 \text{ m/s})^2] = 0.233$ J

Reflect: In the absence of air resistance an object released from rest has a speed of 7.00 m/s after it has fallen 2.50 m. The speed of the ball is much less than this because of the negative work done by friction.

7.59. Set Up: The energy of the box at the edge of the roof is given by: $E_{\text{mech,f}} = E_{\text{mech,i}} - f_k s$. Setting $y_f = 0$ at this point, $y_i = (4.25 \text{ m}) \sin 36° = 2.50$ m. Furthermore, by substituting $K_i = 0$ and $K_f = \frac{1}{2} m v_f^2$ into the conservation equation, $\frac{1}{2} m v_f^2 = m g y_i - f_k s$ or

$$v_f = \sqrt{2 g y_i - 2 f_k s g / w} = \sqrt{2 g (y_i - f_k s / w)}$$

Solve: $v_f = \sqrt{2(9.80 \text{ m/s}^2)[(2.50 \text{ m}) - (22.0 \text{ N})(4.25 \text{ m})/(85.0 \text{ N})]} = 5.24$ m/s

Reflect: Friction does negative work and removes mechanical energy from the system. In the absence of friction the final speed of the toolbox would be 7.00 m/s.

7.61. Set Up: Use $E_{\text{mech,f}} = E_{\text{mech,i}} - f_k s$ to solve for f_k. There is no change in height so $U_i = U_f$ and the energy relation becomes $K_f = K_i - f_k s$.

Solve: (a) First solve for the kinetic energy values:

$$K_i = \tfrac{1}{2} m v_i^2 = \tfrac{1}{2}(375 \text{ kg})(4.5 \text{ m/s})^2 = 3800 \text{ J}$$
$$K_f = \tfrac{1}{2} m v_f^2 = \tfrac{1}{2}(375 \text{ kg})(3.0 \text{ m/s})^2 = 1690 \text{ J}$$

Substituting,

$$f_k = \frac{K_i - K_f}{s} = \frac{3800 \text{ J} - 1690 \text{ J}}{7.0 \text{ m}} = 300 \text{ N}$$

(b) (i) $(K_i - K_f)/K_i = [(3800 \text{ J} - 1690 \text{ J})/3800 \text{ J}] \times 100\% = 56\%$
(b) (ii) $(v_f - v_i)/v_i = [(1.5 \text{ m/s})/(4.5 \text{ m/s})] \times 100\% = 33\%$
Reflect: Friction does negative work and reduces the kinetic energy of the object.

7.63 Set Up: For part (a) use $P = \dfrac{W}{\Delta t}$ to solve for W, the energy the bulb uses. Then set this value equal to $\frac{1}{2} m v^2$ for part (b), and solve for the speed. In part (c), equate the W from part (a) to $U_{\text{grav}} = mgh$ and solve for the height.

Solve: (a) $W = P \Delta t = (100 \text{ W})(3600 \text{ s}) = 3.6 \times 10^5$ J

(b) $K = 3.6 \times 10^5$ J so $v = \sqrt{\dfrac{2K}{m}} = \sqrt{\dfrac{2(3.6 \times 10^5 \text{ J})}{70 \text{ kg}}} = 100$ m/s

(c) $U_{\text{grav}} = 3.6 \times 10^5$ J so $h = \dfrac{U_{\text{grav}}}{mg} = \dfrac{3.6 \times 10^5 \text{ J}}{(70 \text{ kg})(9.80 \text{ m/s}^2)} = 520$ m

Reflect: (b) Olympic runners achieve speeds up to approximately 36 m/s, or roughly one third the result calculated. **(c)** The tallest tree on record, a Redwood, stands 364 ft or 110 m, or 4.7 times smaller than the result.

7.67. Set Up: To lift the skiers, the rope must do positive work to counteract the negative work developed by the component of the gravitational force acting on the total number of skiers, $F_{\text{rope}} = Nmg \cos \phi$. Therefore, use the relation:

$$P_{\text{rope}} = F_{\text{rope}} v = +Nmg(\cos \phi) v \quad \text{where} \quad \phi = (90.0° - 15.0°) = 75.0°$$

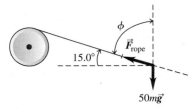

Figure 7.67

Solve:

$$P_{\text{rope}} = F_{\text{rope}} v = [+Nmg(\cos \phi)] v$$

$$= [(50 \text{ riders})(70.0 \text{ kg})(9.80 \text{ m/s}^2)(\cos 75.0°)]\left[(12.0 \text{ km/h})\left(\frac{1 \text{ m/s}}{3.60 \text{ km/h}}\right)\right]$$

$$= 2.96 \times 10^4 \text{ W} = 29.6 \text{ kW}$$

Reflect: Some additional power would be needed to give the riders kinetic energy as they are accelerated from rest.

7.73. Set Up: Find the elapsed time Δt in each case by dividing the distance by the speed, $\Delta t = d/v$. Then calculate the energy as $W = P \Delta t$.

Solve:

$$Running:\ \Delta t = (5.0\ \text{km})/(10\ \text{km/h}) = 0.50\ \text{h} = 1.8 \times 10^3\ \text{s}$$

$$W = (700\ \text{W})(1.8 \times 10^3\ \text{s}) = 1.3 \times 10^6\ \text{J}$$

$$Walking:\ \Delta t = \frac{5.0\ \text{km}}{3.0\ \text{km/h}}\left(\frac{3600\ \text{s}}{\text{h}}\right) = 6.0 \times 10^3\ \text{s}$$

$$W = (290\ \text{W})(6.0 \times 10^3\ \text{s}) = 1.7 \times 10^6\ \text{J}$$

Reflect: The less intense exercise lasts longer.

7.75. Set Up: For part (a), the work performed each day is $W = W_{\text{grav}} = mgh$ where $h = 1.63$ m. The mass of the blood is calculated using the density, $\rho = m/V = 1050\ \text{kg/m}^3$, and the known volume of $V = 7500$ L. The power output is then found from $P = W_{\text{grav}}/t$.

Solve: (a) The work performed in a single day is:

$$W = W_{\text{grav}} = mgh = \rho Vgh = \left(1050\ \frac{\text{kg}}{\text{m}^3}\right)\left(7500\ \text{L} \times \frac{1\ \text{m}^3}{1000\ \text{L}}\right)\left(9.80\ \frac{\text{m}}{\text{s}^2}\right)(1.63\ \text{m}) = 1.26 \times 10^5\ \text{J}$$

(b) $P = \dfrac{W_{\text{grav}}}{t} = \dfrac{(1.26 \times 10^5\ \text{J/day})}{(24\ \text{hr/day})(3600\ \text{s/hr})} = 1.46\ \text{J/s} = 1.46\ \text{W}$

(c) Reflect: The heart actually puts out more power than the value calculated because two forms of energy were not considered; the heart also provides kinetic energy to the blood and generates thermal energy by working against the friction of the vein and vessel walls.

7.77. Set Up: In part (a), the conservation of energy $K_f + U_f = K_i + U_i$ results in $\frac{1}{2}kx_f^2 + 0 = 0 + \frac{1}{2}mv_i^2$ or $k = (mv_i^2)/x^2$. Similarly, the maximum speed may be calculated in part (b) as $v = x\sqrt{k/m}$.

Solve: (a) $k = (mv_i^2)/x^2 = [(1200\ \text{kg})(0.65\ \text{m/s})^2]/[(7.0\ \text{cm})(\text{m}/100\ \text{cm})]^2 = 1.0 \times 10^5\ \text{N/m}$

$$U_{\text{grav}} = K = \tfrac{1}{2}mv^2 = \tfrac{1}{2}(1200\ \text{kg})(0.65\ \text{m/s})^2 = 250\ \text{J}$$

(b) $v = x\sqrt{\dfrac{k}{m}} = (5.0\ \text{cm})\left(\dfrac{\text{m}}{100\ \text{cm}}\right)\sqrt{\dfrac{1.0 \times 10^5\ \text{N/m}}{1200\ \text{kg}}} = 0.46\ \text{m/s}$

Reflect: The maximum permissible speed calculated in (b) is equivalent to roughly 1 mi/h. A practical design for a parking garage should be sufficient for speeds up to 5 mi/h.

7.79. Set Up: Construct a free-body diagram as shown to assess the contributing forces with respect to the path of motion. Note that for circular motion, the velocity vector is always tangent to the circle.

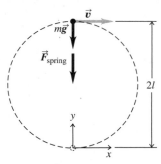

Figure 7.79

Solve: (a) (i) The tension force in the string is always perpendicular to the displacement and thus does no work.

(a) (ii) Considering a complete cycle, $y_2 = y_1 = y$. The work of the gravitational force is $W_{\text{grav}} = -mg(y_2 - y_1) = -mg(y - y) = 0.0$.

(b) (i) The tension does no work.

(b) (ii) Considering a semi-circular portion of the path, $y_2 - y_1 = 2l$ and the work of gravity becomes: $W_{grav} = -mg(y_2 - y_1) = -mg(2l) = -(0.800 \text{ kg})(9.80 \text{ m/s}^2)(2)(1.60 \text{ m}) = -25.1 \text{ J}$.

Reflect: Gravity does negative work as the ball moves upward and positive work as the ball moves downward. For one complete revolution the ball returns to its starting point, there is no change in its gravitational potential energy, and there is no net work done on it by gravity.

7.81. Set Up: Use $K_f + U_f = K_i + U_i - f_k s$ with $f_k = \mu_k m_1 g$, $s = 1.50$ m, and $K_i = 0$. For the 60.0 kg person, who experiences a gravitational potential energy change, let $y_f = 0$ so $y_i = 1.50$ m. Both climbers move with the same speed v. This results in $\frac{1}{2}m_1 v_f^2 + \frac{1}{2}m_2 v_f^2 = m_2 g y_i - \mu_k m_1 g d$, where $m_1 = 75.0$ kg and $m_2 = 60.0$ kg.

Solve: $v_f = \sqrt{\dfrac{2gd(m_2 - \mu_k m_1)}{m_1 + m_2}} = \sqrt{\dfrac{2(9.80 \text{ m/s}^2)(1.50 \text{ m})[60.0 \text{ kg} - 0.250(75.0 \text{ kg})]}{75.0 \text{ kg} + 60.0 \text{ kg}}} = 3.00 \text{ m/s}$

Reflect: Friction does negative work and removes mechanical energy from the system. The tension does positive work on one person and the same magnitude of negative work on the other person and therefore does no net work on the system.

7.83. Set Up: For parts (a) and (b), consider the relations $U_{el} = \frac{1}{2}kx^2$, $K = \frac{1}{2}mv^2$ and $K_f + U_f = K_i + U_i$. In parts (c) and (d), use $|F_{sp}| = kx$. Also recall that $F_{net} = ma$ for (d).

Solve: **(a)** Identical springs are compressed the same distance, so U_{el} is the same for both springs.

(b) The potential energy U_{el} initially stored in each spring is converted to kinetic energy. Since the initial potential energy of the two springs are identical, each object acquires the same kinetic energy.

(c) Solving the kinetic energy relationship for speed, $v = \sqrt{2K/m}$. As both objects have the same kinetic energy K and the 50 g object has smaller mass m, the lighter object has greater speed as it leaves the spring.

(d) Since the spring constant and compressed distance is the same for both springs, $|F_{sp}| = kx$ is also the same for both.

(e) Since $F_{net} = ma = |F_{sp}|$, we can relate the acceleration to the object mass as $a = |F_{sp}|/m$. The 50 g object thus has a greater initial acceleration.

Reflect: The same force produces different accelerations for different masses. The objects both travel the same distance while in contact with the spring, so the object with the greater acceleration reaches a greater speed.

7.85. Set Up: At any point, the potential energy is given by $mgl(1 - \cos\theta)$ where l is the string length and θ is the angle the string makes with the vertical. At the top of the swing, the kinetic energy is zero while the potential energy is a maximum since $\theta = \theta_{max} = 45°$; whereas, at the bottom of travel, the kinetic energy becomes a maximum equal to the U_{grav} of the top of the swing. In terms of conservation of energy, $K_f = U_{grav,i}$ or $\frac{1}{2}mv_f^2 = mgl(1 - \cos 45°)$ where "f" refers to the vertical position of the pendulum and "i" refers to top of travel.

Solve: Solving for the velocity,

$$v_f = \sqrt{2gl(1 - \cos 45°)} = \sqrt{2(9.80 \text{ m/s}^2)(0.80 \text{ m})(1 - \cos 45°)} = 2.1 \text{ m/s}$$

The tension in the string at this instant is found using a force balance of the weight and radial acceleration,

$$T = mg + \frac{mv_f^2}{l} = mg + \frac{2mgl(1 - \cos 45°)}{l} = mg[1 + 2(1 - \cos 45°)]$$

$$T = (0.12 \text{ kg})(9.80 \text{ m/s}^2)(1.5858) = 1.9 \text{ N}$$

Reflect: As the string passes through the vertical the ball has an upward acceleration, due to its circular motion. Therefore, there is a net upward force and the tension is greater than the weight of the ball.

7.87. Set Up: The water is both raised and given speed, so the mechanical energy relation may be applied: $E_{mech} = mgy + \frac{1}{2}mv^2$. One liter of water has a mass of 1 kg.

Solve: $W = E_{mech,f} - E_{mech,i} = mg(y_f - y_i) + \frac{1}{2}m(v_f^2 - v_i^2)$
$= (750 \text{ kg})(9.80 \text{ m/s}^2)(0 - (-14.0 \text{ m})) + \frac{1}{2}(750 \text{ kg})[(18.0 \text{ m/s})^2 - (0.0 \text{ m/s})^2]$
$= 2.24 \times 10^5 \text{ J}$

Reflect: Most of the work done on the water by the pump goes to increasing the kinetic energy of the water.

7.93. Set Up: First set the final kinetic energy of the object to one half of the initial elastic potential energy: $K_f = \frac{1}{2}U_{el,i}$ or $\frac{1}{2}mv_f^2 = \frac{1}{2}(\frac{1}{2}kx_i^2)$. Solving for the speed, $v_f = x_i\sqrt{k/2m}$.

Solve: $v_f = (0.120\text{ m})\sqrt{\dfrac{(35\text{ N/cm})(100\text{ cm/m})}{2(0.250\text{ kg})}} = 10\text{ m/s}$

Reflect: $U_{el,f} = \frac{1}{2}U_{el,i}$. This gives $x_f^2 = \frac{1}{2}x_i^2$ and $x_f = x_i/\sqrt{2} = 0.71x_i$. The object has moved 3.5 cm, less than half the total distance of 12.0 cm.

7.97. Set Up: At the top of the loop, the net force needed to produce the centripetal acceleration, $F_{net} = mv^2/R$, must be equal to or greater than the gravitational force for the car to remain on the track. Thus, a force balance results in $F_{net} = mv^2/R = mg$ or $v = \sqrt{gR}$. Next calculate the height, h, necessary to produce this speed by applying conservation of energy, $U_A + K_A = U_B + K_B$.

Solve: $U_A + K_A = U_B + K_B \quad \Rightarrow \quad mgh_A + 0 = mgh_B + \frac{1}{2}mv_B^2$

$$v_B = \sqrt{gR} \quad \Rightarrow \quad h_A = \frac{mg(2R) + \frac{1}{2}mgR}{mg} = \frac{5}{2}R$$

The height must be at least $2.5R = 2.5(20.0\text{ m}) = 50.0\text{ m}$.

Reflect: If $h < 2.5R$ the car is moving too slowly at B to stay on the track. In that case,

$$\frac{v^2}{R} < g,$$

gravity provides more downward force than is needed to maintain the circular path, and the car falls out of the circle. If $h > 2.5R$, then at B more downward force than gravity alone is needed and this additional downward force is supplied by the normal force that the track exerts on the car.

7.99. Set Up: For part (a), use the conservation of energy relation for nonconservative forces: $K_{foot} + U_{foot} + W_{nc} = K_{top} + U_{top}$. For part (b), compare the maximum static friction force, $f_s = \mu_s mg\cos\theta$, to the skier's weight component down the slope.

Solve: (a) Conservation of energy results in $\frac{1}{2}mv_{foot}^2 - \mu_k(mg\cos\theta)s = mgh$, where $s = h/\sin\theta$. Substituting and solving for the height,

$$\frac{1}{2}mv_{foot}^2 - \mu_k(mg\cos\theta)\left[\frac{h}{\sin\theta}\right] = mgh$$

$$\frac{v_{foot}^2}{2} = gh + \left[\frac{\mu_k gh(\cos\theta)}{\sin\theta}\right]$$

$$h = \frac{v_{foot}^2}{2g\{1 + [(\mu_k\cos\theta)/\sin\theta]\}}$$

$$h = \frac{(15\text{ m/s})^2}{2(9.80\text{ m/s})\{1 + [(0.25\cos40.0°)/\sin40.0°]\}} = 8.8\text{ m}$$

(b) The skier will begin to slide down the hill if the downhill component of her weight is greater than the maximum value of the static friction acting on her. So compare these two quantities.

Maximum static friction: $f_s = \mu_s mg\cos\theta = (0.75)(68\text{ kg})(9.80\text{ m/s}^2)\cos(40.0°) = 380\text{ N}$

Downhill weight component: $w_{g,down} = mg\sin\theta = (68\text{ kg})(9.80\text{ m/s}^2)(\sin40.0°) = 430\text{ N}$

Since 430 N > 380 N, the skier will slide down the hill.

Reflect: As the skier moves up the hill part of her initial kinetic energy is converted to gravitational potential energy by the negative work done by gravity and the rest is removed by the negative work done by friction.

7.101. Set Up: The bag has mass $m = w/g = 17.86$ kg. Use conservation of energy to calculate the speed v of the bag at the top of the cliff: $K_f = U_{el,i}$ or $\frac{1}{2}mv_f^2 = \frac{1}{2}kx^2$ results in $v_f = x\sqrt{k/m}$. Then analyze the projectile motion of the bag: use the vertical motion equation, $y_f = y_i + v_{iy}t + \frac{1}{2}a_y t^2$, to find the time spent in the air; then calculate the horizontal displacement x_f using $x_f = v_{i,x}t$.

Solve: *Speed:* $v_f = x\sqrt{\dfrac{k}{m}} = (0.100\text{ m})\sqrt{\dfrac{(385.0\text{ N/cm})(100\text{ cm/m})}{17.86\text{ kg}}} = 4.64\text{ m/s}$

Projectile motion: $v_{iy} = 0$, $a_y = -9.80\text{ m/s}^2$, $a_x = 0$, $y_i = 0$, $y_f = -3.45\text{ m}$

$$y_f = \tfrac{1}{2}a_y t^2: \qquad t = \sqrt{\dfrac{2y_f}{a_y}} = \sqrt{\dfrac{2(-3.45\text{ m})}{-9.80\text{ m/s}^2}} = 0.839\text{ s}$$

$$a_x = 0: \qquad x_f = v_{ix}t = (4.64\text{ m/s})(0.839\text{ s}) = 3.89\text{ m}$$

$$v_{fx} = v_{ix} = 4.64\text{ m/s}: \qquad v_{fy} = v_{iy} + a_y t = 0 + (-9.80\text{ m/s}^2)(0.839\text{ s}) = -8.22\text{ m/s}$$

$$v_f = \sqrt{v_{fx}^2 + v_{fy}^2} = 9.44\text{ m/s}$$

Reflect: The final speed can also be found using conservation of energy: $\frac{1}{2}kx^2 + mgh = \frac{1}{2}mv_f^2$.

7.105. Set Up: We are given the energy produced per gallon, $0.15E/\text{gal}$ where $E = 1.3 \times 10^8$ J, and wish to find the number of gallons, N. We therefore need to find the kinetic energy used during each acceleration, $N = \Delta K \times (\text{gal}/0.15E)$. Subsequently, for part (b), simply divide the 1 gal by N.

Solve: (a) $\Delta K = \frac{1}{2}m(v_f^2 - v_i^2) = \frac{1}{2}(1500\text{ kg})[(37\text{ m/s}^2) - 0^2] = 1.027 \times 10^6$ J/acceleration

The number of gallons used per acceleration is then,

$$N = \dfrac{\text{gal}}{\text{acceleration}} = \dfrac{\Delta K}{\text{acceleration}} \times \dfrac{\text{gal}}{0.15E} = \dfrac{1.027 \times 10^6\text{ J}}{\text{acceleration}} \times \dfrac{\text{gal}}{0.15(1.3 \times 10^8\text{ J})} = 5.3 \times 10^{-2}\dfrac{\text{gal}}{\text{acceleration}}$$

(c) The number of accelerations is $(1\text{ gal})/(5.3 \times 10^{-2}\text{ gal/acceleration}) = 19$ accelerations.

Reflect: The time of 10.0 s was not used in the calculation. The kinetic energy of the car depends on its speed and not on the magnitude of the acceleration used to achieve that speed.

7.111. Set Up: Use conservation of energy to find the speed of the sled at the edge of the cliff. Let $y_i = 0$ so $y_f = h = 11.0$ m. $K_f + U_f = K_i + U_i$ gives $\frac{1}{2}mv_f^2 + mgh = \frac{1}{2}mv_i^2$ or $v_f = \sqrt{v_i^2 - 2gh}$. Then analyze the projectile motion of the sled: use the vertical component of motion to find the time t that the sled is in the air; then use the horizontal component of the motion with $a_x = 0$ to find the horizontal displacement.

Solve: $v_f = \sqrt{(22.5\text{ m/s})^2 - 2(9.80\text{ m/s}^2)(11.0\text{ m})} = 17.1\text{ m/s}$

$$y_f = v_{i,y}t + \tfrac{1}{2}a_y t^2: \qquad t = \sqrt{\dfrac{2y_f}{a_y}} = \sqrt{\dfrac{2(-11.0\text{ m})}{-9.80\text{ m/s}^2}} = 1.50\text{ s}$$

$$x_f = v_{i,x}t + \tfrac{1}{2}a_x t^2: \qquad x_f = v_{i,x}t = (17.1\text{ m/s})(1.50\text{ s}) = 25.6\text{ m}$$

Reflect: Conservation of energy can be used to find the speed of the sled at any point of the motion but does not specify how far the sled travels while it is in the air.

MOMENTUM

Problems 3, 5, 7, 11, 13, 15, 17, 19, 25, 27, 29, 31, 35, 37, 39, 41, 45, 49, 55, 57, 61, 63, 67, 69, 71, 77, 81, 83, 85, 87

Solutions to Problems

8.3. Set Up: The signs of the velocity components indicate their directions.
Solve: (a) $P_x = p_{Ax} + p_{Cx} = 0 + (10.0 \text{ kg})(-3.0 \text{ m/s}) = -30 \text{ kg} \cdot \text{m/s}$

$$P_y = p_{Ay} + p_{Cy} = (5.0 \text{ kg})(-11.0 \text{ m/s}) + 0 = -55 \text{ kg} \cdot \text{m/s}$$

(b) $P_x = p_{Bx} + p_{Cx} = (6.0 \text{ kg})(10.0 \text{ m/s} \cos 60°) + (10.0 \text{ kg})(-3.0 \text{ m/s}) = 0$

$$P_y = p_{By} + p_{Cy} = (6.0 \text{ kg})(10.0 \text{ m/s} \sin 60°) + 0 = 52 \text{ kg} \cdot \text{m/s}$$

(c) $P_x = p_{Ax} + p_{Bx} + p_{Cx} = 0 + (6.0 \text{ kg})(10.0 \text{ m/s} \cos 60°) + (10.0 \text{ kg})(-3.0 \text{ m/s}) = 0$

$$P_y = p_{Ay} + p_{By} + p_{Cy} = (5.0 \text{ kg})(-11.0 \text{ m/s}) + (6.0 \text{ kg})(10.0 \text{ m/s} \sin 60°) + 0 = -3.0 \text{ kg} \cdot \text{m/s}$$

Reflect: A has no x-component of momentum so P_x is the same in (b) and (c). C has no y-component of momentum so P_y in (c) is the sum of P_y in (a) and (b).

8.5. Set Up: $145 \text{ g} = 0.145 \text{ kg}$; $57 \text{ g} = 0.057 \text{ kg}$
Solve: (a) $p = mv = (0.145 \text{ kg})(45 \text{ m/s}) = 6.5 \text{ kg} \cdot \text{m/s}$

$$K = \tfrac{1}{2}mv^2 = \tfrac{1}{2}(0.145 \text{ kg})(45 \text{ m/s})^2 = 150 \text{ J}$$

(b) (i) $K = \tfrac{1}{2}mv^2$ so $v = \sqrt{\dfrac{2K}{m}} = \sqrt{\dfrac{2(150 \text{ J})}{0.057 \text{ kg}}} = 73 \text{ m/s}$

(ii) $p = mv$ so $v = \dfrac{p}{m} = \dfrac{6.5 \text{ kg} \cdot \text{m/s}}{0.057 \text{ kg}} = 110 \text{ m/s}$

Reflect: K depends on v^2 and p depends on v so a smaller increase in v is required for the lighter ball to have the same kinetic energy.

8.7. Set Up: $p = mv$. From Problem 8.6, $p = \sqrt{2mK}$.
Solve: (a) $p = m_c v_c$. $p_d = m_d v_d$. $v_d = v_c$ and $m_d = 3m_c$ so $p_d = 3m_c v_c = 3p$
(b) $p = \sqrt{2m_c K_c}$. $p_d = \sqrt{2m_d K_d}$. $K_d = K_c$ and $m_d = 3m_c$ so $p_d = \sqrt{2(3m_c)K_c} = \sqrt{3}p$

Reflect: In (b), $v_c = \dfrac{p}{m_c}$ and $v_d = \dfrac{p_d}{m_d} = \dfrac{\sqrt{3}p}{3m_c} = \dfrac{v_c}{\sqrt{3}}$. When the mass is increased the speed must be decreased to keep the same kinetic energy.

8.11. Set Up: Use coordinates where $+x$ is in the direction of the motion of the ball before it is caught. Figure 8.11 gives before and after sketches for the system of ball plus player. $(v_{b,i})_x = 100$ mph $= 44.7$ m/s

Before   After

Figure 8.11

Solve: There is no external force in the x-direction, so $P_{i,x} = P_{f,x}$. This gives $m_b v_{b,i} = (m_b + m_p)v_f$.

$$v_f = \left(\frac{m_b}{m_b + m_p}\right)v_{b,i} = \left(\frac{0.145 \text{ kg}}{0.145 \text{ kg} + 65 \text{ kg}}\right)(44.7 \text{ m/s}) = 9.9 \text{ cm/s}$$

Reflect: There is a large increase in the mass that is moving so to maintain constant total momentum there is a large decrease in speed.

8.13. Set Up: Let $+x$ be to the right.
Solve: (a) $P_{i,x} = P_{f,x}$ says $(0.250)v_{A,i} = (0.250 \text{ kg})(-0.120 \text{ m/s}) + (0.350 \text{ kg})(0.650 \text{ m/s})$ and $v_{A,i} = 0.790$ m/s
(b) $K_i = \frac{1}{2}(0.250 \text{ kg})(0.790 \text{ m/s})^2 = 0.0780$ J

$$K_f = \frac{1}{2}(0.250 \text{ kg})(0.120 \text{ m/s})^2 + \frac{1}{2}(0.350 \text{ kg})(0.650 \text{ m/s})^2 = 0.0757 \text{ J}$$

and $\Delta K = K_f - K_i = -0.0023$ J.
Reflect: The total momentum of the system is conserved but the total kinetic energy decreases.

8.15. Set Up: Let $+x$ be east. Momentum is conserved in the east-west direction. Momentum is not conserved in the north-south direction because of the horizontal external force exerted by the tracks.
Solve: (a) The 25.0 kg mass retains its eastward velocity and momentum, the velocity of the handcar doesn't change. The final velocity of the handcar is 5.00 m/s, east.
(b) The mass has velocity zero relative to the earth. $P_{i,x} = P_{f,x}$ says $(200.0 \text{ kg})(5.00 \text{ m/s}) = (175.0 \text{ kg})v_{f,x}$ and $v_{f,x} = 5.71$ m/s. The final velocity of the handcar is 5.71 m/s, east.
(c) $P_{i,x} = P_{f,x}$ says $(200.0 \text{ kg})(5.00 \text{ m/s}) + (25.0 \text{ kg})(-6.00 \text{ m/s}) = (225.0 \text{ kg})v_{f,x}$ and $v_{f,x} = 3.78$ m/s. The final velocity of the handcar is 3.78 m/s, east.
(d) As in part (a) the 25.0 kg mass retains its eastward velocity and momentum. The velocity of the handcar doesn't change. The final velocity of the handcar is 5.00 m/s, east.
Reflect: By Newton's third law, the force exerted on the handcar by the mass that is thrown is opposite in direction to the force on the mass. In parts (a) and (d), these forces aren't parallel to the tracks and don't change the speed of the handcar. In part (b), the force on the mass is west so the reaction force on the handcar is east and the handcar speeds up. In part (c), the handcar exerts a force on the mass that is east, so the reaction force on the handcar is west and the handcar slows down.

8.17. Set Up: Use coordinates where $+x$ is in the direction of motion of the bullet. Figure 8.17 gives before and after sketches for the system of bullet and block.

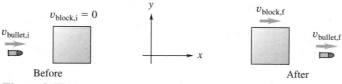

Before   After

Figure 8.17

Solve: There is no external force in the x direction, so $P_{i,x} = P_{f,x}$. This gives $m_{bullet}v_{bullet,i} = m_{bullet}v_{bullet,f} + m_{block}v_{block,f}$.

$$v_{block,f} = \left(\frac{m_{bullet}}{m_{block}}\right)(v_{bullet,i} - v_{bullet,f}) = \left(\frac{4.25 \times 10^{-3} \text{ kg}}{1.12 \text{ kg}}\right)(375 \text{ m/s} - 122 \text{ m/s}) = 0.960 \text{ m/s}$$

Reflect: The momentum lost by the bullet equals the momentum gained by the block.

8.19. Set Up: Use coordinates where $+x$ is to the right and $+y$ is upward.
Solve: *Collision:* There is no external horizontal force during the collision so $P_{i,x} = P_{f,x}$. This gives $(5.00 \text{ kg})(12.0 \text{ m/s}) = (10.0 \text{ kg})v_f$ and $v_f = 6.0 \text{ m/s}$.
Motion after the collision: Only gravity does work and the initial kinetic energy of the combined chunks is converted entirely to gravitational potential energy when the chunk reaches its maximum height h above the valley floor. Conservation of energy gives

$$\tfrac{1}{2}m_{tot}v^2 = m_{tot}gh \text{ and } h = \frac{v^2}{2g} = \frac{(6.0 \text{ m/s})^2}{2(9.8 \text{ m/s}^2)} = 1.8 \text{ m}$$

Reflect: After the collision the energy of the system is $\tfrac{1}{2}m_{tot}v^2 = \tfrac{1}{2}(10.0 \text{ kg})(6.0 \text{ m/s})^2 = 180$ J when it is all kinetic energy and it is $m_{tot}gh = (10.0 \text{ kg})(9.8 \text{ m/s}^2)(1.8 \text{ m}) = 180$ J when it is all gravitational potential energy. Mechanical energy is conserved during the motion after the collision. But before the collision the total energy of the system is $\tfrac{1}{2}(5.0 \text{ kg})(12.0 \text{ m/s})^2 = 360$ J; 50% of the mechanical energy is dissipated during the inelastic collision of the two chunks.

8.25. Set Up: Apply conservation of momentum to the collision and the work-energy equation to the motion after the collision. Let $+x$ be the direction the bullet was traveling before the collision.
Solve: *Motion after the collision:* $W_{other} + U_i + K_i = K_f + U_f$. There is no change in height so $U_i = U_f$. The block stops so $K_f = 0$. $n = mg$, $f_k = \mu_k mg$ and $W_{other} = W_f = -\mu_k mgs$. Let V be the speed of the block immediately after the collision, so $K_i = \tfrac{1}{2}mV^2$. Therefore, $-\mu_k mgs + \tfrac{1}{2}mV^2 = 0$ and

$$V = \sqrt{2\mu_k gs} = \sqrt{2(0.20)(9.80 \text{ m/s}^2)(0.230 \text{ m})} = 0.950 \text{ m/s}.$$

Collision: $m_{bullet}v_{bullet,i} = mV$, where $m = 1.20 \text{ kg} + 5.00 \times 10^{-3} \text{ kg} = 1.205 \text{ kg}$.

$$v_{bullet,i} = \frac{(1.205 \text{ kg})(0.950 \text{ m/s})}{5.00 \times 10^{-3} \text{ kg}} = 229 \text{ m/s}$$

Reflect: Our analysis neglects the friction force exerted by the surface on the block during the collision, since we assume momentum is conserved in the collision. The collision forces between the bullet and block are much larger than the friction force, so this approximation is justified.

8.27. Set Up: Use coordinates where $+x$ is east and $+y$ is south. Let the big fish be A and the small fish be B. The system of the two fish before and after the collision is sketched in Figure 8.27.

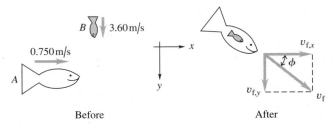

Before   After

Figure 8.27

Solve: There are no external forces on the fish so $P_{i,x} = P_{f,x}$ and $P_{i,y} = P_{f,y}$.

$$P_{i,x} = P_{f,x} \text{ gives } (11.5 \text{ kg})(0.750 \text{ m/s}) = (12.75 \text{ kg})v_{f,x} \text{ so } v_{f,x} = 0.676 \text{ m/s}$$

$$P_{i,y} = P_{f,y} \text{ gives } (1.25 \text{ kg})(3.60 \text{ m/s}) = (12.75 \text{ kg})v_{f,y} \text{ so } v_{f,y} = 0.353 \text{ m/s}$$

$v = \sqrt{v_{f,x}^2 + v_{f,y}^2} = 0.763 \text{ m/s}$. $v_{f,x} = v_f\cos\phi$ so $\cos\phi = \dfrac{0.676 \text{ m/s}}{0.763 \text{ m/s}}$ and $\phi = 27.6°$. The large fish has velocity 0.763 m/s in a direction 27.6° south of east.

Reflect: Momentum is a vector and we must treat each component separately.

8.29. Set Up: Use coordinates where $+x$ is east and $+y$ is south. The system of two cars before and after the collision is sketched in Figure 8.29. Neglect friction from the road during the collision. The enmeshed cars have mass 2000 kg + 1500 kg = 3500 kg.

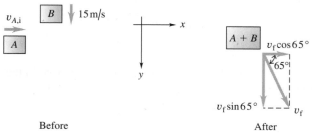

Before   After

Figure 8.29

Solve: There are no external horizontal forces during the collision, so $P_{i,x} = P_{f,x}$ and $P_{i,y} = P_{f,y}$.
(a) $P_{i,x} = P_{f,x}$ gives $(1500 \text{ kg})(15 \text{ m/s}) = (3500 \text{ kg})v_f\sin65°$ and $v_f = 7.1 \text{ m/s}$
(b) $P_{i,y} = P_{f,y}$ gives $(2000 \text{ kg})v_{A,i} = (3500 \text{ kg})v_f\cos65°$. And then with $v_f = 7.1 \text{ m/s}$, $v_{A,i} = 5.2 \text{ m/s}$.
Reflect: Momentum is a vector and we must treat each component separately.

8.31. Set Up: For an elastic collision with B initially stationary, the final velocities are

$$v_A = \left(\frac{m_A - m_B}{m_A + m_B}\right)v \text{ and } v_B = \left(\frac{2m_A}{m_A + m_B}\right)v.$$

Apply these equations with $m_A = 0.300 \text{ kg}$, $m_B = 0.150 \text{ kg}$ and $v = 0.80 \text{ m/s}$.
Solve: (a) $v_A = \left(\dfrac{0.300 \text{ kg} - 0.150 \text{ kg}}{0.300 \text{ kg} + 0.150 \text{ kg}}\right)(0.80 \text{ m/s}) = 0.27 \text{ m/s}$,

$$v_B = \left(\frac{2[0.300 \text{ kg}]}{0.300 \text{ kg} + 0.150 \text{ kg}}\right)(0.80 \text{ m/s}) = 1.07 \text{ m/s}.$$

The 0.300 kg glider moves to the right at 0.27 m/s and the 0.150 kg glider moves to the right at 1.07 m/s.
(b) $K_{A,f} = \frac{1}{2}m_A v_A^2 = \frac{1}{2}(0.300 \text{ kg})(0.27 \text{ m/s})^2 = 0.011 \text{ J}$;

$$K_{B,f} = \tfrac{1}{2}m_B v_B^2 = \tfrac{1}{2}(0.150 \text{ kg})(1.07 \text{ m/s})^2 = 0.086 \text{ J}$$

Reflect: The relative velocity after the collision has magnitude 1.07 m/s − 0.27 m/s = 0.80 m/s, the same as before the collision. The initial kinetic energy of the system is $K_{A,i} = \frac{1}{2}(0.300 \text{ kg})(0.80 \text{ m/s})^2 = 0.096 \text{ J}$. The final kinetic energy is $K_{A,f} + K_{B,f} = 0.097 \text{ J}$, the same as the initial kinetic energy, apart from a slight difference due to rounding.

8.35. Set Up: Let A be the bowling ball and let B be the table-tennis ball. $m_A \gg m_B$.

Solve: $v_A = \left(\dfrac{m_A - m_B}{m_A + m_B}\right)V \approx \dfrac{m_A}{m_A}V = V; \ v_B = \left(\dfrac{2m_A}{m_A + m_B}\right)V = \dfrac{2m_A}{m_A}V = 2V$

The bowling ball continues to move at speed V in the same direction and the table-tennis ball moves in that same direction with speed $2V$.

Reflect: The table-tennis ball gains some kinetic energy so the bowling ball must slow down slightly, but not much because of its large mass.

8.37. Set Up: Assume the ball is initially moving to the right, and let this be the $+x$ direction. The ball stops, so its final velocity is zero.

Solve: (a) $J_x = mv_{f,x} - mv_{i,x} = 0 - (0.145 \text{ kg})(36.0 \text{ m/s}) = -5.22 \text{ kg} \cdot \text{m/s}$ The magnitude of the impulse applied to the ball is $5.22 \text{ kg} \cdot \text{m/s}$.

(b) $J_x = F_x \Delta t$ so $F_x = \dfrac{J_x}{\Delta t} = \dfrac{-5.22 \text{ kg} \cdot \text{m/s}}{20.0 \times 10^{-3} \text{ s}} = -261 \text{ N}$

Reflect: The signs of J_x and F_x show that both these quantities are to the left.

8.39. Set Up: (a) Take the $+x$ direction to be along the final direction of motion of the ball. The initial speed of the ball is zero. **(b)** Take the $+x$ direction to be in the direction the ball is traveling before it is hit by the opponent.

Solve: (a) $J_x = mv_{f,x} - mv_{i,x} = (57 \times 10^{-3} \text{ kg})(73.14 \text{ m/s} - 0) = 4.2 \text{ kg} \cdot \text{m/s}$

$$F_x = \dfrac{J_x}{\Delta t} = \dfrac{4.2 \text{ kg} \cdot \text{m/s}}{30.0 \times 10^{-3} \text{ s}} = 140 \text{ N}$$

(b) $J_x = mv_{f,x} - mv_{i,x} = (57 \times 10^{-3} \text{ kg})(-55 \text{ m/s} - 73.14 \text{ m/s}) = -7.3 \text{ kg} \cdot \text{m/s}$

$$F_x = \dfrac{J_x}{\Delta t} = \dfrac{-7.3 \text{ kg} \cdot \text{m/s}}{30.0 \times 10^{-3} \text{ s}} = -240 \text{ N}$$

Reflect: The signs of J_x and F_x show their direction. $140 \text{ N} = 31 \text{ lb}$. This very attainable force has a large effect on the light ball. 140 N is 250 times the weight of the ball.

8.41. Set Up: The impulse is the area under the F_x versus t curve between t_i and t_f. Let the $+x$ direction be to the left, so F_x is positive.

Solve: (a) $J_x = \text{area} = F_x \Delta t = (4.0 \times 10^3 \text{ N})(5.0 \times 10^{-3} \text{ s} - 2.0 \times 10^{-3} \text{ s}) = 12.0 \text{ N} \cdot \text{s}$
J_x is positive so the impulse is to the left.

(b) $J_x = mv_{f,x} - mv_{i,x}$

(i) $v_{i,x} = 0$, so $v_{f,x} = \dfrac{J_x}{m} = \dfrac{12.0 \text{ N} \cdot \text{s}}{0.150 \text{ kg}} = 80 \text{ m/s}$.

The final velocity of the ball is 80 m/s, to the left.

(ii) $v_{i,x} = -30.0 \text{ m/s}$, so $v_{f,x} = v_{i,x} + \dfrac{J_x}{m} = -30.0 \text{ m/s} + 80.0 \text{ m/s} = 50.0 \text{ m/s}$

The final velocity of the ball is 50 m/s, to the left.

Reflect: In both (i) and (ii) of part (b) the change in velocity is the same, 80 m/s.

8.45. Set Up: Represent each atom as a point. Use coordinates with the origin at the carbon atom and with the oxygen atom on the $+x$ axis.

Solve: $x_{\text{cm}} = \dfrac{m_C x_C + m_O x_O}{m_C + m_O}$. $x_C = 0$ and $x_O = 0.111 \text{ nm}$, so

$$x_{\text{cm}} = \dfrac{16(0.111 \text{ nm})}{12 + 16} = 0.063 \text{ nm}.$$

The center of mass is along the line connecting the two atoms, 0.063 nm from the carbon atom and 0.048 nm from the oxygen atom.

Reflect: The center of mass is closer to the more massive oxygen atom.

8.49. Set Up: Use coordinates with the origin at the 20 N weight and with $+y$ toward the 10 N weight and $+x$ toward the 40 N weight. Since weight and mass are proportional, we can use weight in place of mass in calculating the coordinates of the center of mass.

Solve: (a) $x_{cm} = \dfrac{(20\,\text{N})(0) + (10\,\text{N})(0) + (30\,\text{N})(1.00\,\text{m}) + (40\,\text{N})(1.00\,\text{m})}{20\,\text{N} + 10\,\text{N} + 30\,\text{N} + 40\,\text{N}} = 0.70\,\text{m}$

$y_{cm} = \dfrac{(20\,\text{N})(0) + (10\,\text{N})(1.00\,\text{m}) + (30\,\text{N})(1.00\,\text{m}) + (40\,\text{N})(0)}{20\,\text{N} + 10\,\text{N} + 30\,\text{N} + 40\,\text{N}} = 0.40\,\text{m}$

The center of mass is 0.40 m above and 0.70 m to the right of the 20 N weight.

(b) The center of mass of the four rods taken together is at the center of the square. Including their weight moves the center of mass of the complete system closer to the center of the square, so moves the center of mass calculated in (a) up and to the left.

Reflect: The object is not symmetric and its center of mass is not at its geometric center.

8.55. Set Up: Use Eq. 8.26. $\dfrac{\Delta m}{\Delta t} = -\dfrac{m}{160}$. Solve for v_{ex}.

Solve: $a = -\dfrac{v_{ex}}{m}\dfrac{\Delta m}{\Delta t}$.

$$v_{ex} = -\dfrac{a}{\left(\dfrac{\Delta m}{\Delta t}/m\right)} = \dfrac{-15.0\,\text{m/s}^2}{\left(-\dfrac{m}{160}\right)/m} = 2.40 \times 10^3\,\text{m/s} = 2.40\,\text{km/s}$$

Reflect: The acceleration is proportional to the speed of the exhaust gas and to the rate at which mass is ejected.

8.57. Set Up: Initially $m = 2.04 \times 10^6\,\text{kg}$ and $F = 35 \times 10^6\,\text{N}$. Since the shuttle is at the earth's surface, the gravity force on it must be included. The free-body diagram for the shuttle at liftoff is shown in Figure 8.57. Take $+y$ to be upward.

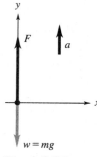

Figure 8.57

Solve: (a) $\Sigma F_y = ma_y$ gives $F - mg = ma$ and

$$a = \frac{F}{m} - g = \frac{35 \times 10^6\,\text{N}}{2.04 \times 10^6\,\text{kg}} - 9.80\,\text{m/s}^2 = 7.4\,\text{m/s}^2$$

(b) $\dfrac{\Delta m}{\Delta t} = -\dfrac{230{,}000\,\text{kg}}{60\,\text{s}} = -3.8 \times 10^3\,\text{kg/s}$

(c) $F = -v_{ex}\dfrac{\Delta m}{\Delta t}$ so

$$v_{ex} = -\frac{F}{\Delta m/\Delta t} = -\frac{35 \times 10^6\,\text{N}}{-3.8 \times 10^3\,\text{kg/s}} = 9.2 \times 10^3\,\text{m/s} = 9.2\,\text{km/s}$$

Reflect: In part (a) it is important to include the gravity force on the shuttle; the thrust is only 1.75 times its weight.

8.61. Set Up: Let $+x$ be the initial direction of motion of the bullet and let $+y$ be the direction of the final motion of the bullet. $m_A = 2.50 \times 10^{-3}$ kg (bullet); $m_B = 0.100$ kg (stone). The before and after diagrams are sketched in Figure 8.61.

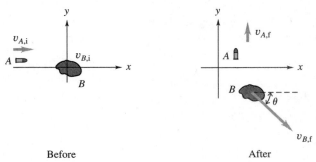

Before   After

Figure 8.61

Solve: **(a)** $P_{i,x} = P_{f,x}$ says $m_A v_{A,i} = m_B v_{B,f,x}$. $(2.50 \times 10^{-3}$ kg$)(500$ m/s$) = (0.100$ kg$)v_{B,f,x}$ and $v_{B,f,x} = 12.5$ m/s. $P_{i,y} = P_{f,y}$ says $0 = m_A v_{A,f,y} + m_B v_{B,f,y}$. $0 = (2.50 \times 10^{-3}$ kg$)(300$ m/s$) + (0.100$ kg$)v_{B,f,y}$ and $v_{B,f,y} = -7.5$ m/s. Then

$$v_{B,f} = \sqrt{(v_{B,f,x})^2 + (v_{B,f,y})^2} = \sqrt{(12.5 \text{ m/s})^2 + (-7.5 \text{ m/s})^2} = 14.6 \text{ m/s}.$$

$$\tan\theta = \frac{7.5 \text{ m/s}}{12.5 \text{ m/s}}. \ \theta = 31.0°.$$

The stone rebounds at 14.6 m/s in a direction of 31.0° from the initial direction of motion of the bullet.
(b) $K_i = \frac{1}{2}(2.50 \times 10^{-3}$ kg$)(500$ m/s$)^2 = 312$ J and

$$K_f = \frac{1}{2}(2.50 \times 10^{-3} \text{ kg})(300 \text{ m/s})^2 + \frac{1}{2}(0.100 \text{ kg})(14.6 \text{ m/s})^2 = 123 \text{ J}.$$

$K_f < K_i$ and the collision is inelastic.
Reflect: Each component of momentum is separately conserved. Kinetic energy is a scalar, not a vector, and we never consider its "components".

8.63. Set Up: Let $+x$ be the direction the neutron is traveling initially. Since the collision is head-on all the motion is along the x axis. $m_A = 1.67 \times 10^{-27}$ kg. $m_B = 1.66 \times 10^{-26}$ kg. For a perfectly elastic collision with object B originally at rest, the final speeds are given by Eqs. 8.11 and 8.12.
Solve: **(a)** $m_A v_{A,i,x} = (m_A + m_B)v_{f,x}$. $(1.67 \times 10^{-27}$ kg$)(45.0$ km/s$) = (1.827 \times 10^{-26}$ kg$)v_{f,x}$ and $v_{f,x} = 4.11$ km/s

$$K_i = \frac{1}{2}(1.67 \times 10^{-27} \text{ kg})(45.0 \times 10^3 \text{ m/s})^2 = 1.69 \times 10^{-18} \text{ J}$$

$$K_f = \frac{1}{2}(1.827 \times 10^{-26} \text{ kg})(4.11 \times 10^3 \text{ m/s})^2 = 1.54 \times 10^{-19} \text{ J}.$$

$$\frac{K_f}{K_i} = \frac{1.54 \times 10^{-19} \text{ J}}{1.69 \times 10^{-18} \text{ J}} = 0.0911.$$

(b) $v_{A,f} = \left(\dfrac{m_A - m_B}{m_A + m_B}\right)v_{A,i} = \left(\dfrac{1.67 \times 10^{-27} \text{ kg} - 1.66 \times 10^{-26} \text{ kg}}{1.67 \times 10^{-27} \text{ kg} + 1.66 \times 10^{-26} \text{ kg}}\right)(45.0 \text{ km/s}) = -36.8 \text{ km/s}$

$$v_{B,f} = \left(\frac{2m_A}{m_A + m_B}\right)v_{A,i} = \left(\frac{2[1.67 \times 10^{-27} \text{ kg}]}{1.67 \times 10^{-27} \text{ kg} + 1.66 \times 10^{-26} \text{ kg}}\right)(45.0 \text{ km/s}) = 8.23 \text{ km/s}$$

$$K_{f,B} = \frac{1}{2}(1.66 \times 10^{-26} \text{ kg})(8.23 \times 10^3 \text{ m/s})^2 = 5.62 \times 10^{-19} \text{ J}.$$

$$\frac{K_{f,B}}{K_{i,A}} = \frac{5.62 \times 10^{-19} \text{ J}}{1.69 \times 10^{-18} \text{ J}} = 0.333$$

and 33.3% of the neutron's original kinetic energy is transferred to the boron nucleus.

Reflect: In an elastic collision with object B initially at rest, the fraction of the initial kinetic energy of object A that is transferred to B depends on the ratio of the masses of A and B. When A and B have equal masses, all the kinetic energy of A is transferred to B.

8.67. Set Up: Let the system be the two masses and the spring. The system is sketched in Figure 8.67, in its initial and final situations. Use coordinates where $+x$ is to the right. Call the masses A and B.

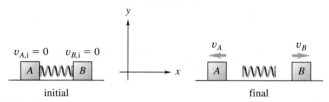

initial   final

Figure 8.67

Solve: $P_{i,x} = P_{f,x}$ so $0 = (1.50 \text{ kg})(-v_A) + (1.50 \text{ kg})(v_B)$ and $v_A = v_B$. Energy conservation says the potential energy originally stored in the spring is all converted into kinetic energy of the blocks, so $\frac{1}{2}kx_i^2 = \frac{1}{2}mv_A^2 + \frac{1}{2}mv_B^2$. $v_A = v_B$ so

$$v_A = x_i\sqrt{\frac{k}{2m}} = (0.200 \text{ m})\sqrt{\frac{175 \text{ N/m}}{2[1.50 \text{ kg}]}} = 1.53 \text{ m/s}$$

Reflect: If the objects have different masses they will end up with different speeds. The lighter one will have the greater speed, since they end up with equal magnitudes of momentum.

8.69. Set Up: Let $+x$ be to the right. Apply conservation of momentum to the collision and conservation of energy to the motion of the block after the collision. Let the bullet be A and the block be B. Let V be the velocity of the block just after the collision.

Solve: *Motion of block after the collision:* $K_i = U_{grav,f}$. $\frac{1}{2}m_BV^2 = m_Bgh$.

$$V = \sqrt{2gh} = \sqrt{2(9.80 \text{ m/s}^2)(0.450 \times 10^{-2} \text{ m})} = 0.297 \text{ m/s}$$

Collision: $v_{B,f} = 0.297 \text{ m/s}$. $P_{i,x} = P_{f,x}$ gives $m_Av_{A,i} = m_Av_{A,f} + m_Bv_{B,f}$.

$$v_{A,f} = \frac{m_Av_{A,i} - m_Bv_{B,f}}{m_A} = \frac{(5.00 \times 10^{-3} \text{ kg})(450 \text{ m/s}) - (1.00 \text{ kg})(0.297 \text{ m/s})}{5.00 \times 10^{-3} \text{ kg}} = 391 \text{ m/s}$$

Reflect: We assume the block moves very little during the time it takes the bullet to pass through it.

8.71. Set Up: Let $+x$ be to the right. $w_A = 800 \text{ N}$, $w_B = 600 \text{ N}$ and $w_C = 1000 \text{ N}$.

Solve: $P_{i,x} = P_{f,x}$ gives $0 = m_Av_{A,f,x} + m_Bv_{B,f,x} + m_Cv_{C,f,x}$.

$$v_{C,f,x} = \frac{m_Av_{A,f,x} + m_Bv_{B,f,x}}{m_C} = \frac{w_Av_{A,f,x} + w_Bv_{B,f,x}}{w_C}.$$

$$v_{C,f,x} = \frac{(800 \text{ N})(-[5.00 \text{ m/s}]\cos 30.0°) + (600 \text{ N})(+[7.00 \text{ m/s}]\cos 36.9°)}{1000 \text{ N}} = -0.105 \text{ m/s}$$

The sleigh's velocity is 0.105 m/s, to the left.

Reflect: The vertical component of the momentum of the system consisting of the two people and the sleigh is not conserved, because of the net force exerted on the sleigh by the ice while they jump.

8.77. Set Up: Apply conservation of momentum to the collision and conservation of energy to the motion after the collision. Let $+x$ be to the right.

Solve: *Collision:* $P_{i,x} = P_{f,x}$ gives $(4.20 \times 10^{-3} \text{ kg})v = (4.20 \times 10^{-3} \text{ kg} + 2.50 \text{ kg})V$

Motion after the collision: The kinetic energy of the block immediately after the collision is converted entirely to gravitational potential energy at the maximum angle of swing. Figure 8.77 shows that the maximum height h the block swings up is given by $h = (0.750 \text{ m})(1 - \cos 34.7°) = 0.133 \text{ m}$.

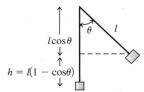

Figure 8.77

Conservation of energy gives $\frac{1}{2}m_{tot}V^2 = m_{tot}gh$ and $V = \sqrt{2gh} = 1.61 \text{ m/s}$
Then the conservation of momentum equation gives

$$v = \left(\frac{4.20 \times 10^{-3} \text{ kg} + 2.50 \text{ kg}}{4.20 \times 10^{-3} \text{ kg}}\right)(1.61 \text{ m/s}) = 960 \text{ m/s}$$

Reflect: The original speed of the bullet is nearly three times the speed of sound in air. Kinetic energy is not conserved in the collision. Our analysis assumes that the block moves very little during the collision, while the bullet is coming to rest relative to the block. This is a good approximation since the velocity the block gets in the collision is much less than the initial velocity of the bullet.

8.81. Set Up: For an elastic collision with object B initially at rest, $v_A = \left(\frac{m_A - m_B}{m_A + m_B}\right)v$ and $v_B = \left(\frac{2m_A}{m_A + m_B}\right)v$. A is the alpha particle and B is the gold nucleus. $v = 1.25 \times 10^7 \text{ m/s}$.

Solve: $v_A = \left(\frac{6.65 \times 10^{-27} \text{ kg} - 3.27 \times 10^{-25} \text{ kg}}{6.65 \times 10^{-27} \text{ kg} + 3.27 \times 10^{-25} \text{ kg}}\right)(1.25 \times 10^7 \text{ m/s}) = -1.20 \times 10^7 \text{ m/s}$

$$v_B = \left(\frac{2(6.65 \times 10^{-27} \text{ kg})}{6.65 \times 10^{-27} \text{ kg} + 3.27 \times 10^{-25} \text{ kg}}\right)(1.25 \times 10^7 \text{ m/s}) = 4.98 \times 10^5 \text{ m/s}$$

Reflect: In an elastic collision of a light object with a stationary heavy object, the light object bounces back with its speed only slightly reduced.

8.83. Set Up: Apply conservation of momentum. Take $+x$ to be to the right. For (a) the collision is elastic. Let A be the cart that is initially moving to the right. For (b) the two carts stick together and the combined object has final speed v_f.

Solve: **(a)** $P_{i,x} = P_{f,x}$ gives $Mv_0 + M(-v_0) = Mv_A + Mv_B$ and $v_A = -v_B$. For an elastic collision $v_{B,f} - v_{A,f} = -(v_{B,i} - v_{A,i})$. This gives $v_B - v_A = -(-v_0 - v_0) = 2v_0$. $v_A = -v_B$ so $v_B = v_0$ and $v_A = -v_0$ A rebounds with speed v_0 and B rebounds with speed v_0.

(b) $P_{i,x} = P_{f,x}$ gives $Mv_0 + M(-v_0) = (2M)v_f$ and $v_f = 0$. Both carts are rest after the collision.

Reflect: Collision (b) is totally inelastic. All the initial kinetic energy is converted to other forms. Any collision in which the objects stick together must be inelastic.

8.85. Set Up: Call the fragments A and B, with $m_A = 2.0$ kg and $m_B = 5.0$ kg. After the explosion fragment A moves in the $+x$ direction with speed v_A and fragment B moves in the $-x$ direction with speed v_B.

Solve: $P_{i,x} = P_{f,x}$ gives $0 = m_A v_A + m_B(-v_B)$ and

$$v_A = \left(\frac{m_B}{m_A}\right)v_B = \left(\frac{5.0 \text{ kg}}{2.0 \text{ kg}}\right)v_B = 2.5v_B.$$

$$\frac{K_A}{K_B} = \frac{\frac{1}{2}m_A v_A^2}{\frac{1}{2}m_B v_B^2} = \frac{\frac{1}{2}(2.0 \text{ kg})(2.5v_B)^2}{\frac{1}{2}(5.0 \text{ kg})v_B^2} = \frac{12.5}{5.0} = 2.5.$$

$K_A = 100$ J so $K_B = 250$ J

Reflect: In an explosion the lighter fragment receives the most of the liberated energy.

8.87. Set Up: Use constant acceleration equations for the free-fall to find the speed of the acorn just before it collides with the bird. The system of the bird and acorn is sketched in Figure 8.87, before and after the collision. Use the coordinates shown. Assume that gravity can be neglected during the collision.

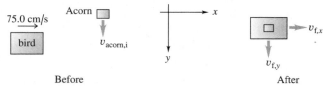

Before   After

Figure 8.87

Solve: **(a)** For the fall of the acorn, $v_{0y} = 0$, $a_y = 9.80$ m/s² and $y - y_0 = 17.5$ cm.

$v_y^2 = v_{0y}^2 + 2a_y(y - y_0)$ and $v_y = \sqrt{2a_y(y - y_0)} = 18.5$ m/s. $v_{\text{acorn},i} = 18.5$ m/s

$P_{i,x} = P_{f,x}$ gives $(135 \text{ g})(75.0 \text{ cm/s}) = (135 \text{ g} + 15.0 \text{ g})v_{f,x}$ and $v_{f,x} = 67.5$ cm/s.

$P_{i,y} = P_{f,y}$ gives $(15.0 \text{ g})(1850 \text{ cm/s}) = (135 \text{ g} + 15.0 \text{ g})v_{f,y}$ and $v_{f,y} = 185$ cm/s.

(b) $v_f = \sqrt{v_{f,x}^2 + v_{f,y}^2} = 197$ cm/s; $\tan\phi = \dfrac{v_{f,x}}{v_{f,y}} = \dfrac{67.5 \text{ cm/s}}{185 \text{ cm/s}}$ so $\phi = 20.0°$

Reflect: Momentum is a vector and there are separate equations for conservation of its x and y components.

ROTATIONAL MOTION

9

Problems 3, 5, 9, 15, 17, 19, 21, 25, 27, 33, 35, 37, 41, 43, 47, 49, 51, 53, 55, 57, 59, 65, 69, 71, 73

Solutions to Problems

9.3. Set Up: For one revolution, $\Delta\theta = 2\pi$ rad. Assume constant angular velocity, so

$$\omega = \frac{\Delta\theta}{\Delta t}.$$

The second hand makes 1 revolution in 1 minute = 60.0 s. The minute hand makes 1 revolution in 1 h = 3600 s, and the hour hand makes 1 revolution in 12 h = 43,200 s.

Solve: (a) *second hand* $\omega = \dfrac{2\pi \text{ rad}}{60 \text{ s}} = 0.105 \text{ rad/s}$;

minute hand $\omega = \dfrac{2\pi \text{ rad}}{3600 \text{ s}} = 1.75 \times 10^{-3} \text{ rad/s}$;

hour hand $\omega = \dfrac{2\pi \text{ rad}}{43,200 \text{ s}} = 1.45 \times 10^{-4} \text{ rad/s}$

(b) The period is the time for 1 revolution. Second hand, 1min; minute hand, 1 h; hour hand, 12 h.
Reflect: When the angular velocity is constant, $\omega = \omega_{av}$.

9.5. Set Up: 1 rev = 2π rad
Solve: (a) The period is the time for 1 rev, so period = 2.25 s.
$\qquad\qquad\qquad\qquad\qquad\qquad\qquad$ **(b)** $\omega = \dfrac{\theta}{t} = \dfrac{2\pi \text{ rad}}{2.25 \text{ s}} = 2.79 \text{ rad/s}$

Reflect: The angular speed and period don't depend on the radius of the wheel.

9.9. Set Up: 1 rpm = $(2\pi/60)$ rad/s. Period $T = \dfrac{2\pi}{\omega}$. $\theta - \theta_0 = \omega t$.

Solve: (a) $\omega = (1900)(2\pi \text{ rad}/60 \text{ s}) = 199 \text{ rad/s}$

(b) $35° = (35°)(\pi/180°) = 0.611$ rad. $t = \dfrac{\theta - \theta_0}{\omega} = \dfrac{0.611 \text{ rad}}{199 \text{ rad/s}} = 3.1 \times 10^{-3} \text{ s}$

(c) $(18 \text{ rad/s})\left(\dfrac{60 \text{ s}}{1 \text{ min}}\right)\left(\dfrac{1 \text{ rev}}{2\pi \text{ rad}}\right) = 172 \text{ rpm}$

(d) For the propeller rotating at 1900 rpm, $T = \dfrac{2\pi \text{ rad}}{\omega} = \dfrac{2\pi \text{ rad}}{199 \text{ rad/s}} = 0.032 \text{ s}$.

Reflect: The period is inversely proportional to the angular velocity.

9.15. Set Up: Let the direction the wheel is rotating be the positive sense of rotation. Since the wheel starts from rest, $\omega_0 = 0$. $1 \text{ rev} = 2\pi \text{ rad}$

Solve: (a) $\omega = \omega_0 + \omega_0 t$ gives $t = \dfrac{\omega - \omega_0}{\alpha} = \dfrac{8.00 \text{ rad/s} - 0}{0.640 \text{ rad/s}^2} = 12.5 \text{ s}$

(b) $\omega^2 = \omega_0^2 + 2\alpha(\theta - \theta_0)$ gives $\theta - \theta_0 = \dfrac{\omega^2 - \omega_0^2}{2\alpha} = \dfrac{(8.00 \text{ rad/s})^2 - 0}{2(0.640 \text{ rad/s}^2)} = 50.0 \text{ rad} = 7.96 \text{ rev}.$

Reflect: We could also do part (b) by

$$\theta - \theta_0 = \left(\frac{\omega_0 + \omega}{2}\right)t = \left(\frac{0 + 8.00 \text{ rad/s}}{2}\right)(12.5 \text{ s}) = 50.0 \text{ rad}.$$

9.17. Set Up: $500.0 \text{ rpm} = 8.33 \text{ rev/s}$. Let the direction of rotation of the flywheel be positive.

Solve: (a) $(\theta - \theta_0) = \left(\dfrac{\omega_0 + \omega}{2}\right)t$ gives

$$\omega = \frac{2(\theta - \theta_0)}{t} - \omega_0 = \frac{2(200.0 \text{ rev})}{30.0 \text{ s}} - 8.33 \text{ rev/s} = 5.00 \text{ rev/s} = 300 \text{ rpm}$$

(b) Use information in part (a) to find α: $\omega = \omega_0 + \omega_0 t$ gives

$$\alpha = \frac{\omega - \omega_0}{t} = \frac{5.00 \text{ rev/s} - 8.33 \text{ rev/s}}{30.0 \text{ s}} = -0.111 \text{ rev/s}^2.$$

Then, with $\omega = 0$ and $\omega_0 = 8.33 \text{ rev/s}$ the equation $\omega = \omega_0 + \omega_0 t$ gives

$$t = \frac{\omega - \omega_0}{\alpha} = \frac{0 - 8.33 \text{ rev/s}}{-0.111 \text{ rev/s}^2} = 75.0 \text{ s}.$$

$$\theta - \theta_0 = \left(\frac{\omega_0 + \omega}{2}\right)t = \left(\frac{8.33 \text{ rev/s} + 0}{2}\right)(75.0 \text{ s}) = 312 \text{ rev}.$$

Reflect: The angular acceleration is negative because the wheel is slowing down.

9.19. Set Up: Let the direction of rotation of the flywheel be positive.
Solve: $\theta - \theta_0 = \omega_0 t + \frac{1}{2}\alpha t^2$ gives

$$\omega_0 = \frac{\theta - \theta_0}{t} - \frac{1}{2}\alpha t = \frac{60.0 \text{ rad}}{4.00 \text{ s}} - \frac{1}{2}(2.25 \text{ rad/s}^2)(4.00 \text{ s}) = 10.5 \text{ rad/s}.$$

Reflect: At the end of the 4.00 s interval, $\omega = \omega_0 + \alpha t = 19.5 \text{ rad/s}$.

$$\theta - \theta_0 = \left(\frac{\omega_0 + \omega}{2}\right)t = \left(\frac{10.5 \text{ rad/s} + 19.5 \text{ rad/s}}{2}\right)(4.00 \text{ s}) = 60.0 \text{ rad},$$

which checks.

9.21. Set Up: $r = 6.5 \text{ mm} = 6.5 \times 10^{-3} \text{ m}$. $\omega = 1250 \text{ rpm} = 130.9 \text{ rad/s}$
Solve: $v = r\omega = (6.5 \times 10^{-3} \text{ m})(130.9 \text{ rad/s}) = 0.85 \text{ m/s}$
Reflect: $v = r\omega$ requires that ω be in rad/s.

9.25. Set Up: $r_1 = 4.0 \text{ in.}$, $r_2 = 1.5 \text{ in.}$ and $\omega_2 = 75 \text{ rpm}$. The 3 in. sprocket and 24 in. wheel are mounted on the same axle and turn at the same rate.
Solve: All points on the chain have the same speed so $v_1 = v_2$; points on the rim of each sprocket move at the same tangential speed. $v = r\omega$ gives $r_1\omega_1 = r_2\omega_2$.

$$\omega_1 = \left(\frac{r_2}{r_1}\right)\omega_2 = \left(\frac{1.5 \text{ in.}}{4.0 \text{ in.}}\right)(75 \text{ rpm}) = 28 \text{ rpm}$$

Reflect: The large sprocket turns at a slower rate than the small sprocket.

9.27. Set Up: Tape has $v = 1\frac{7}{8}$ in./s $= 4.762$ cm/s. The tape takes 45 min $= 2700$ s to completely unwind.

Solve: (a) $\omega = \dfrac{v}{r} = \dfrac{4.762 \text{ cm/s}}{0.10 \text{ cm}} = 47.6$ rad/s

(b) (i) *full:* $r = 1$ in. and $\omega = \dfrac{v}{r} = \dfrac{1\frac{7}{8} \text{ in./s}}{1 \text{ in.}} = 1\frac{7}{8}$ rad/s $= 1.88$ rad/s

(ii) *empty:* $r = \frac{7}{16}$ in. and $\omega = \dfrac{v}{r} = \dfrac{1\frac{7}{8} \text{ in./s}}{\frac{7}{16} \text{ in.}} = 4.29$ rad/s

(c) The length of the tape is $vt = (4.762 \text{ cm/s})(2700 \text{ s}) = 129$ m.

Reflect: When the radius is smaller the angular velocity must be larger to produce the same tangential speed.

9.33. Set Up: (a) Each mass is a distance

$$\frac{\sqrt{2(0.400 \text{ m})^2}}{2} = \frac{0.400 \text{ m}}{\sqrt{2}}$$

from the axis. **(b)** each mass is 0.200 m from the axis. **(c)** two masses are on the axis and two are

$$\frac{0.400 \text{ m}}{\sqrt{2}}$$

from the axis.

Solve: (a) $I = \sum mr^2 = 4(0.200 \text{ kg})\left(\dfrac{0.400 \text{ m}}{\sqrt{2}}\right)^2 = 0.0640$ kg · m²

(b) $I = 4(0.200 \text{ kg})(0.200 \text{ m})^2 = 0.0320$ kg · m²

(c) $I = 2(0.200 \text{ kg})\left(\dfrac{0.400 \text{ m}}{\sqrt{2}}\right)^2 = 0.0320$ kg · m²

Reflect: The value of I depends on the location of the axis.

9.35. Set Up: (a) For a solid cylinder of mass m, $I = \frac{1}{2}mR^2$ and for a thin-walled hollow cylinder of mass M, $I = MR^2$. **(b)** For a solid sphere of mass M, $I = \frac{2}{5}MR^2$ and for a thin-walled hollow sphere of mass M, $I = \frac{2}{3}MR^2$.

Solve: (a) $2(\frac{1}{2}mR^2) = \frac{1}{2}mR^2 + MR^2$ and $M = m/2$.

(b) Total mass is doubled, so the mass of the foil equals the mass M of the sphere. $I_{\text{old}} = \frac{2}{5}MR^2$.

$$I_{\text{new}} = \frac{2}{5}MR^2 + \frac{2}{3}MR^2 = \frac{16}{15}MR^2. \quad \frac{I_{\text{new}}}{I_{\text{old}}} = \frac{\frac{16}{15}MR^2}{\frac{2}{5}MR^2} = \left(\frac{16}{15}\right)\left(\frac{5}{2}\right) = \frac{8}{3}$$

Reflect: The moment of inertia of the compound object is the algebraic sum of I for each of its pieces.

9.37. Set Up: For the solid sphere $I_{\text{sph}} = \frac{2}{5}M_{\text{sph}}R^2$ and for the solid disk $I_{\text{disk}} = \frac{1}{2}M_{\text{disk}}R^2$. The two objects have the same radius R.

Solve: $I_{\text{sph}} = I_{\text{disk}}$ so $\frac{2}{5}M_{\text{sph}}R^2 = \frac{1}{2}M_{\text{disk}}R^2$. $M_{\text{disk}} = \frac{4}{5}M_{\text{sph}}$ and $w_{\text{disk}} = \frac{4}{5}w_{\text{sph}} = \frac{4}{5}(5.55 \text{ N}) = 4.44$ N

Reflect: The disk has smaller weight because for it I is a larger fraction of MR^2.

9.41. Set Up: The volume of a hollow cylinder of inner radius R_1, outer radius R_2 and length L is $V = \pi L(R_2^2 - R_1^2)$. $I = \frac{1}{2}M(R_1^2 + R_2^2)$. Let the density be ρ; $M = \rho V$.

Solve: (a) $\omega = \dfrac{1 \text{ rev}}{1.75 \text{ s}} = 0.571$ rev/s $= 3.59$ rad/s

$$K = \tfrac{1}{2}I\omega^2 = \tfrac{1}{2}(\tfrac{1}{2}M[R_1^2 + R_2^2])\omega^2 = \tfrac{1}{4}\rho\pi L(R_2^2 - R_1^2)(R_1^2 + R_2^2)\omega^2 = \tfrac{1}{4}\rho\pi L(R_2^4 - R_1^4)\omega^2$$

$$L = \frac{4K}{\rho\pi(R_2^4 - R_1^4)\omega^2} = \frac{4(2.5 \times 10^6 \text{ J})}{(2.20 \times 10^3 \text{ kg/m}^3)\pi([1.50 \text{ m}]^4 - [0.500 \text{ m}]^4)(3.59 \text{ rad/s})^2} = 22.5 \text{ m}$$

(b) Length L is proportional to $1/\omega^2$, so doubling ω reduces L to $(22.5 \text{ m})/4 = 5.63$ m.

Reflect: The length could also be reduced by using material of larger density.

9.43. Set Up: The speed v of the weight is related to ω of the cylinder by $v = R\omega$, where $R = 0.325$ m. Use coordinates where $+y$ is upward and $y_i = 0$ for the weight. $y_f = -h$, where h is the unknown distance the weight descends. Let $m = 1.50$ kg and $M = 3.25$ kg. For the cylinder $I = \frac{1}{2}MR^2$.

Solve: (a) Conservation of energy says $K_i + U_i = K_f + U_f$. $K_i = 0$ and $U_i = 0$. $U_f = mgy_f = -mgh$.

$$K_f = \tfrac{1}{2}mv^2 + \tfrac{1}{2}I\omega^2 = \tfrac{1}{2}mv^2 + \tfrac{1}{2}\left(\tfrac{1}{2}MR^2\right)\left(\frac{v}{R}\right)^2 = \left(\tfrac{1}{2}m + \tfrac{1}{4}M\right)v^2$$

$$\left(\tfrac{1}{2}m + \tfrac{1}{4}M\right)v^2 - mgh = 0$$

$$h = \frac{\left(\tfrac{1}{2}m + \tfrac{1}{4}M\right)v^2}{mg} = \frac{\left[\tfrac{1}{2}(1.50\ \text{kg}) + \tfrac{1}{4}(3.25\ \text{kg})\right](2.50\ \text{m/s})^2}{(1.50\ \text{kg})(9.80\ \text{m/s}^2)} = 0.664\ \text{m}$$

(b) $\omega = \dfrac{v}{R} = \dfrac{2.50\ \text{m/s}}{0.325\ \text{m}} = 7.69\ \text{rad/s}$

Reflect: The net work done by the rope that connects the cylinder and weight is zero. The speed v of the weight equals the tangential speed at the outer surface of the cylinder, and this gives $v = R\omega$.

9.47. Set Up: $I = \frac{2}{5}MR^2$. R can be calculated from the circumference c, $c = 2\pi R$. $\omega = 25.0$ rev/s $= 157$ rad/s. For part (c), $v_{cm} = R\omega$.

Solve: (a) $K_{tot} = K_{cm} + K_{rot}$ with $K_{cm} = \frac{1}{2}Mv_{cm}^2$ and $K_{rot} = \frac{1}{2}I_{cm}\omega^2$. $K_{cm} = \frac{1}{2}(0.145\ \text{kg})(38.0\ \text{m/s})^2 = 105$ J

$$R = \frac{c}{2\pi} = \frac{0.230\ \text{m}}{2\pi} = 0.0366\ \text{m and}$$

$$K_{rot} = \tfrac{1}{2}\left(\tfrac{2}{5}MR^2\right)\omega^2 = \tfrac{1}{5}(0.145\ \text{kg})(0.0366\ \text{m})^2(157\ \text{rad/s})^2 = 0.958\ \text{J}$$

$$K_{tot} = 105\ \text{J} + 0.958\ \text{J} = 106\ \text{J}$$

(b) $\dfrac{K_{rot}}{K_{tot}} = \dfrac{0.958\ \text{J}}{106\ \text{J}} = 0.904\%$

(c) $K_{tot} = K_{cm} + K_{rot}$. $K_{cm} = \frac{1}{2}Mv_{cm}^2$. $v_{cm} = R\omega$ and $I_{cm} = \frac{2}{5}MR^2$, so

$$K_{rot} = \tfrac{1}{2}I_{cm}\omega^2 = \tfrac{1}{2}\left(\tfrac{2}{5}MR^2\right)\left(\frac{v_{cm}}{R}\right)^2 = \tfrac{1}{5}Mv_{cm}^2.$$

$$K_{tot} = \tfrac{1}{2}Mv_{cm}^2 + \tfrac{1}{5}Mv_{cm}^2 = \tfrac{7}{10}Mv_{cm}^2.$$

$$\frac{K_{rot}}{K_{tot}} = \frac{\tfrac{1}{5}Mv_{cm}^2}{\tfrac{7}{10}Mv_{cm}^2} = \tfrac{1}{5}\left(\tfrac{10}{7}\right) = \tfrac{2}{7} = 28.6\%$$

(d) A tennis ball can be modeled as a thin-walled hollow sphere, so $I_2 = \frac{2}{3}MR^2$. The baseball had $I_1 = \frac{2}{5}MR^2$ so $I_2/I_1 = \frac{5}{3}$. I for the tennis ball is larger by a factor of $\frac{5}{3}$. For the tennis ball $K_{rot} = \frac{5}{3}(0.958\ \text{J}) = 1.60\ \text{J}$ and $K_{tot} = 107\ \text{J}$.

$$\frac{K_{rot}}{K_{tot}} = 1.50\%.$$

Reflect: For the same M and R the mass of the tennis ball is distributed farther from the axis and I is larger. Note that $v_{cm} = R\omega$ does *not* apply in parts (a) and (b) since the ball is not rolling. There is no fixed relation between v_{cm} and ω for a thrown ball.

9.49. Set Up: The ball has moment of inertia $I_{cm} = \frac{2}{3}mR^2$. Rolling without slipping means $v_{cm} = R\omega$. Use coordinates where $+y$ is upward and $y = 0$ at the bottom of the hill, so $y_i = 0$ and $y_f = h = 5.00$ m.

Solve: (a) Conservation of energy gives $K_i + U_i = K_f + U_f$. $U_i = 0$, $K_f = 0$ (stops). Therefore $K_i = U_f$ and $\frac{1}{2}mv_{cm}^2 + \frac{1}{2}I_{cm}\omega^2 = mgh$.

$$\frac{1}{2}I_{cm}\omega^2 = \frac{1}{2}\left(\frac{2}{3}mR^2\right)\left(\frac{v_{cm}}{R}\right)^2 = \frac{1}{3}mv_{cm}^2 \text{ so } \frac{5}{6}mv_{cm}^2 = mgh.$$

$$v_{cm} = \sqrt{\frac{6gh}{5}} = \sqrt{\frac{6(9.80 \text{ m/s}^2)(5.00 \text{ m})}{5}} = 7.67 \text{ m/s}$$

and

$$\omega = \frac{v_{cm}}{R} = \frac{7.67 \text{ m/s}}{0.113 \text{ m}} = 67.9 \text{ rad/s}.$$

(b) $K_{rot} = \frac{1}{2}I\omega^2 = \frac{1}{3}mv_{cm}^2 = \frac{1}{3}(0.426 \text{ kg})(7.67 \text{ m/s})^2 = 8.35$ J

Reflect: Its translational kinetic energy at the base of the hill is $\frac{1}{2}mv_{cm}^2 = \frac{3}{2}K_{rot} = 12.52$ J. Its total kinetic energy is 20.9 J. This equals its final potential energy: $mgh = (0.426 \text{ kg})(9.80 \text{ m/s}^2)(5.00 \text{ m}) = 20.9$ J

9.51. Set Up: The solid cylinder has $I_{cm} = \frac{1}{2}mR^2$ and the solid sphere has $I_{cm} = \frac{2}{5}mR^2$. $v_{cm} = R\omega$ for each. Use coordinates where $+y$ is upward and $y = 0$ at the bottom of the hill. Then $y_i = 0$ and $y_f = h$ for each object. Let the maximum heights h be h_{cyl} and h_{sph}.

Solve: (a) Conservation of energy gives $K_i + U_i = K_f + U_f$. $K_f = 0$, $U_i = 0$ so $K_i = U_f$.

$$mgh = \frac{1}{2}mv_{cm}^2 + \frac{1}{2}I_{cm}\omega^2$$

cylinder: $\frac{1}{2}I_{cm}\omega^2 = \frac{1}{2}\left(\frac{1}{2}mR^2\right)\left(\frac{v_{cm}}{R}\right)^2 = \frac{1}{4}mv_{cm}^2$

$\frac{3}{4}mv_{cm}^2 = mgh$ and $h_{cyl} = \dfrac{3v_{cm}^2}{4g} = \dfrac{3(6.5 \text{ m/s})^2}{4(9.80 \text{ m/s}^2)} = 3.23$ m

sphere: $\frac{1}{2}I_{cm}\omega^2 = \frac{1}{2}\left(\frac{2}{5}mR^2\right)\left(\frac{v_{cm}}{R}\right)^2 = \frac{1}{5}mv_{cm}^2$

$\frac{7}{10}mv_{cm}^2 = mgh$ and $h_{sph} = \dfrac{7v_{cm}^2}{10g} = \dfrac{7(6.5 \text{ m/s})^2}{10(9.80 \text{ m/s}^2)} = 3.02$ m

(b) The cylinder has a larger I_{cm} and hence more initial kinetic energy.

(c) The final expression for h does not involve either m or R so the answers would not change.

Reflect: The total kinetic energy, translational plus rotational, has been converted to potential energy when the object momentarily comes to rest at its maximum height.

9.53. Set Up: The hollow cylinder has $I = \frac{1}{2}m(R_1^2 + R_2^2)$, where $R_1 = 0.200$ m and $R_2 = 0.350$ m. Use coordinates where $+y$ is upward and $y = 0$ at the initial position of the cylinder. Then $y_i = 0$ and $y_f = -d$, where d is the distance it has fallen. $v_{cm} = R\omega$.

Solve: (a) Conservation of energy gives $K_i + U_i = K_f + U_f$. $K_i = 0$, $U_i = 0$. $0 = U_f + K_f$ and $0 = -mgd + \frac{1}{2}mv_{cm}^2 + \frac{1}{2}I_{cm}\omega^2$

$$\frac{1}{2}I\omega^2 = \frac{1}{2}\left(\frac{1}{2}m[R_1^2 + R_2^2]\right)(v_{cm}/R_2)^2 = \frac{1}{4}m[1 + (R_1/R_2)^2]v_{cm}^2,$$

so $\frac{1}{2}\left(1 + \frac{1}{2}[1 + (R_1/R_2)^2]\right)v_{cm}^2 = gd$ and

$$d = \frac{\left(1 + \frac{1}{2}[1 + (R_1/R_2)^2]\right)v_{cm}^2}{2g} = \frac{(1 + 0.663)(6.66 \text{ m/s})^2}{2(9.80 \text{ m/s}^2)} = 3.76 \text{ m}$$

(b) $K_f = \frac{1}{2}mv_{cm}^2$; no rotation

$$mgd = \frac{1}{2}mv_{cm}^2 \text{ and } v_{cm} = \sqrt{2gd} = \sqrt{2(9.80 \text{ m/s}^2)(3.76 \text{ m})} = 8.58 \text{ m/s}$$

(c) In part (a) the cylinder has rotational as well as translational kinetic energy and therefore less translational speed at a given kinetic energy. The kinetic energy comes from a decrease in gravitational potential energy and that is the same, so in (a) the translational speed is less.

Reflect: If part (a) were repeated for a solid cylinder, $R_1 = 0$ and $d = 3.39$ m. For a thin-walled hollow cylinder, $R_1 = R_2$ and $d = 4.52$ cm. Note that all of these answers are independent of the mass m of the cylinder.

9.55. Set Up: The linear acceleration a of the elevator equals the tangential acceleration of a point on the rim of the shaft. $a = 0.150g = 1.47$ m/s. For the shaft, $R = 0.0800$ m.

Solve: $a_{\text{tan}} = R\alpha$ so $a = R\alpha$ and $\alpha = \dfrac{1.47 \text{ m/s}^2}{0.0800 \text{ m}} = 18.4 \text{ rad/s}^2$

Reflect: In $a_{\text{tan}} = R\alpha$, α is in units of rad/s^2.

9.57. Set Up: The distance d the car travels equals the arc length s traveled by a point on the rim of the tire, so $d = r\theta$. The odometer reading depends on the angle through which the wheels have turned.

Solve: (a) $d = 0.10$ mi $= 528$ ft. $r = 12$ in. $= 1$ ft. $\theta = \dfrac{d}{r} = \dfrac{528 \text{ ft}}{1 \text{ ft}} = 528$ rad $= 84.0$ rev.

(b) $d = r\theta$. 5000 rev $= 3.14 \times 10^4$ rad. $d = (1 \text{ ft})(3.14 \times 10^4 \text{ rad}) = 3.14 \times 10^4$ ft $= 5.95$ mi

(c) With proper 24 in. diameter tires the angular displacement for $d = 500$ mi $= 2.64 \times 10^6$ ft is

$$\theta = \frac{d}{r} = \frac{2.64 \times 10^6 \text{ ft}}{1 \text{ ft}} = 2.64 \times 10^6 \text{ rad}.$$

With 28 in. tires this θ corresponds to $d = r\theta = \left(\frac{14}{12} \text{ ft}\right)(2.64 \times 10^6 \text{ rad}) = 3.08 \times 10^6$ ft $= 583$ mi.

Reflect: In $s = r\theta$ the angle θ must be expressed in radians.

9.59. Set Up: $U = Mgy_{\text{cm}}$. Let $y_i = 0$, so $U_i = 0$ and $y_f = -0.500$ m. $I = \frac{1}{3}ML^2$. $K_i = 0$. $W_{\text{other}} = 0$, so conservation of energy gives $K_i + U_i = K_f + U_f$. $v = r\omega$.

Solve: (a) $U_f - U_i = Mgy_{\text{cm}} - 0 = (0.160 \text{ kg})(9.80 \text{ m/s}^2)(-0.500 \text{ m}) = -0.784$ J
(b) $K_f + U_f = 0. \frac{1}{2}I\omega^2 + Mgy_f = 0. \frac{1}{6}ML^2\omega^2 + Mgy_f = 0.$

$$\omega = \frac{\sqrt{-6gy_f}}{1.00 \text{ m}} = \frac{\sqrt{-6(9.80 \text{ m/s}^2)(-0.500 \text{ m})}}{1.00 \text{ m}} = 5.42 \text{ rad/s}$$

(c) $v = r\omega = (1.00 \text{ m})(5.42 \text{ rad/s}) = 5.42$ m/s
(d) $\frac{1}{2}mv^2 = mgy$ and $v = \sqrt{2gy} = \sqrt{2(9.80 \text{ m/s}^2)(1.00 \text{ m})} = 4.43$ m/s
Reflect: The end of the stick falls a vertical distance of 1.00 m but has a final speed greater than for a particle that falls freely 1.00 m.

9.65. Set Up: My total mass is $m = 90$ kg. I model my head as a uniform sphere of radius 8 cm. I model my trunk and legs as a uniform solid cylinder of radius 12 cm. I model my arms as slender rods of length 60 cm. $\omega = 72$ rev/min $= 7.5$ rad/s.
Solve: (a) $I_{\text{tot}} = \frac{2}{5}(0.070m)(0.080 \text{ m})^2 + \frac{1}{2}(0.80m)(0.12 \text{ m})^2 + 2(\frac{1}{3})(0.13m)(0.60 \text{ m})^2 = 3.3 \text{ kg} \cdot \text{m}^2$
(b) $K_{\text{rot}} = \frac{1}{2}I\omega^2 = \frac{1}{2}(3.3 \text{ kg} \cdot \text{m}^2)(7.5 \text{ rad/s})^2 = 93$ J
Reflect: According to these estimates about 85% of the total I is due to the outstretched arms. If the initial translational kinetic energy $\frac{1}{2}mv^2$ of the skater is converted to this rotational kinetic energy as he goes into a spin, his initial speed must be 1.4 m/s.

9.69. Set Up: $I_{\text{cm}} = \frac{2}{5}mR^2$. If the stone rolls without slipping, $v_{\text{cm}} = R\omega$. If there is no friction the stone slides without rolling and has no rotational kinetic energy. Use coordinates where $+y$ is upward and $y = 0$ at the bottom of the hill.
Solve: After the stone is launched into the air its translational kinetic energy is converted to potential energy. At its maximum height h, $mgh = \frac{1}{2}mv_{\text{cm}}^2$, where v_{cm} is its translational speed at the bottom of the hill.

(a) Apply conservation of energy to the motion down the hill: $mgH = \frac{1}{2}mv_{cm}^2$ and $v_{cm}^2 = 2gH$. Then $mgh = \frac{1}{2}mv_{cm}^2 = \frac{1}{2}m(2gH)$ and $h = H$.

(b) Now the initial potential energy is converted to both translational and rotational kinetic energy as the stone rolls down the hill:

$$mgH = \frac{1}{2}mv_{cm}^2 + \frac{1}{2}I_{cm}\omega^2$$

$$\frac{1}{2}I_{cm}\omega^2 = \frac{1}{2}\left(\frac{2}{5}mR^2\right)\left(\frac{v_{cm}}{R}\right)^2 = \frac{1}{5}mv_{cm}^2 \text{ so } mgH = \frac{7}{10}mv_{cm}^2 \text{ and } v_{cm}^2 = \frac{10}{7}gH$$

This gives $mgh = \frac{1}{2}m\left(\frac{10}{7}gH\right)$ and $h = \frac{5}{7}H$.

(c) In (a) all the initial gravitational potential energy is converted back to final gravitational potential energy; the kinetic energy at the final maximum height is zero. In (b) the stone gains rotational kinetic energy as it rolls down the hill and it still has this rotational kinetic energy at its maximum height; not all the initial potential energy is converted into the final potential energy.

Reflect: Both answers do not depend on the mass or radius of the stone. But the answer to (b) depends on how the mass is distributed; the answer would be different for a hollow rolling sphere.

9.71. Set Up: For a uniform bar pivoted about one end, $I = \frac{1}{3}mL^2$. $v = 5.0\text{ km/h} = 1.4\text{ m/s}$.

(a) $60° = (\pi/3)$ rad. The average angular speed of each arm and leg is $\dfrac{\pi/3 \text{ rad}}{1 \text{ s}} = 1.05\text{ rad/s}$.

(b) $I = \frac{1}{3}m_{arm}L_{arm}^2 + \frac{1}{3}m_{leg}L_{leg}^2 = \frac{1}{3}\left[(0.13)(75\text{ kg})(0.70\text{ m})^2 + (0.37)(75\text{ kg})(0.90\text{ m})^2\right]$

$$I = 9.08\text{ kg}\cdot\text{m}^2$$

$$K_{rot} = \frac{1}{2}I\omega^2 = \frac{1}{2}(9.08\text{ kg}\cdot\text{m}^2)(1.05\text{ rad/s})^2 = 5.0\text{ J}$$

(c) $K_{tran} = \frac{1}{2}mv^2 = \frac{1}{2}(75\text{ kg})(1.4\text{ m/s})^2 = 73.5\text{ J}$ and $K_{tot} = K_{tran} + K_{rot} = 78.5\text{ J}$

(d) $\dfrac{K_{rot}}{K_{tran}} = \dfrac{5.0\text{ J}}{78.5\text{ J}} = 6.4\%$

Reflect: If you swing your arms more vigorously more of your energy input goes into the kinetic energy of walking and it is more effective exercise.

9.73. Set Up: Use projectile motion to find the speed v the marble needs at the edge of the pit to make it to the level ground on the other side. The marble must travel 36 m horizontally while falling vertically 20 m. Let $+y$ be downward. Then use conservation of energy to relate the initial height to the speed v at the edge of the pit. $W_{other} = 0$ so conservation of energy gives $K_i + U_i = K_f + U_f$. Let $y_f = 0$ so $U_f = 0$ and $y_i = h$. $K_i = 0$. Rolling without slipping means $v = R\omega$. $K = \frac{1}{2}I_{cm}\omega^2 + \frac{1}{2}mv^2 = \frac{1}{2}\left(\frac{2}{5}mR^2\right)\omega^2 + \frac{1}{2}mv^2 = \frac{7}{10}mv^2$.

Solve: (a) Projectile motion: $v_{0y} = 0$. $a_y = 9.80\text{ m/s}^2$. $y - y_0 = 20\text{ m}$. $y - y_0 = v_{0y}t + \frac{1}{2}a_yt^2$ gives

$$t = \sqrt{\frac{2(y - y_0)}{a_y}} = 2.02\text{ s}.$$

Then $x - x_0 = v_{0x}t$ gives

$$v = v_{0x} = \frac{x - x_0}{t} = \frac{36\text{ m}}{2.02\text{ s}} = 17.8\text{ m/s}.$$

Motion down the hill: $U_i = K_f$. $mgh = \frac{7}{10}mv^2$. $h = \dfrac{7v^2}{10g} = \dfrac{7(17.8\text{ m/s})^2}{10(9.80\text{ m/s}^2)} = 22.6\text{ m}$.

(b) $\frac{1}{2}I\omega^2 = \frac{1}{5}mv^2$, independent of R. I is proportional to R^2 but ω^2 is proportional to $1/R^2$ for a given translational speed v.

Reflect: The answer to part (a) also does not depend on the mass of the marble. But, it does depend on how the mass is distributed within the object. The answer would be different if the object were a hollow spherical shell.

DYNAMICS OF ROTATIONAL MOTION

Problems 1, 5, 9, 11, 13, 15, 19, 21, 25, 27, 31, 33, 39, 43, 45, 47, 51, 53, 57, 61, 63, 65, 67, 69, 75, 79, 81, 83

Solutions to Problems

10.1. Set Up: Let counterclockwise torques be positive. $\tau = Fl$ with $l = r\sin\phi$.
Solve: **(a)** $\tau = +(10.0\text{ N})(4.00\text{ m})\sin 90.0° = 40.0\text{ N}\cdot\text{m}$, counterclockwise.
(b) $\tau = +(10.0\text{ N})(4.00\text{ m})\sin 60.0° = 34.6\text{ N}\cdot\text{m}$, counterclockwise.
(c) $\tau = +(10.0\text{ N})(4.00\text{ m})\sin 30.0° = 20.0\text{ N}\cdot\text{m}$, counterclockwise.
(d) $\tau = -(10.0\text{ N})(2.00\text{ m})\sin 60.0° = -17.3\text{ N}\cdot\text{m}$, clockwise.
(e) $\tau = 0$ since the force acts on the axis and $l = 0$
(f) $\tau = 0$ since the line of action of the force passes through the location of the axis and $l = 0$.
Reflect: The torque of a force depends on the direction of the force and where it is applied to the object.

10.5. Set Up: Let counterclockwise torques be positive. $\tau = Fl$
Solve: $\tau_1 = -F_1 l_1 = -(18.0\text{ N})(0.090\text{ m}) = -1.62\text{ N}\cdot\text{m}$. $\tau_2 = +F_2 l_2 = +(26.0\text{ N})(0.090\text{ m}) = +2.34\text{ N}\cdot\text{m}$.
$\tau_3 = +F_3 l_3 = +(14.0\text{ N})(0.127\text{ m}) = +1.78\text{ N}\cdot\text{m}$.
$\Sigma\tau = \tau_1 + \tau_2 + \tau_3 = 2.50\text{ N}\cdot\text{m}$, counterclockwise.
Reflect: It is important to take into account the direction of each torque when computing the net torque.

10.9. Set Up: Let the direction the grindstone is rotating be positive. From Table 9.2, $I = \frac{1}{2}mR^2$. Use the information about the motion to calculate α. In $\Sigma\tau = I\alpha$, α must be in rad/s^2.
Solve: $\omega_0 = 850\text{ rev/min} = 89.0\text{ rad/s}$. $\omega = 0$. $t = 7.50\text{ s}$. $\omega = \omega_0 + \alpha t$ gives

$$\alpha = \frac{\omega - \omega_0}{t} = \frac{0 - 89.0\text{ rad/s}}{7.50\text{ s}} = -11.9\text{ rad/s}^2.$$

The net torque is due to the kinetic friction force f_k between the ax and the grindstone. $\Sigma\tau = I\alpha$ gives $-f_k R = \left(\frac{1}{2}mR^2\right)\alpha$ and

$$f_k = \frac{-mR\alpha}{2} = -\frac{(50.0\text{ kg})(0.260\text{ m})(-11.9\text{ rad/s}^2)}{2} = 77.4\text{ N}.$$

$f_k = \mu_k n$ so $\mu_k = \frac{f_k}{n} = \frac{77.4\text{ N}}{160\text{ N}} = 0.484$
Reflect: $\Sigma\tau = I\alpha$ relates the forces on the rotating object to the motion of the object.

10.11. Set Up: For the pulley $I = \frac{1}{2}M(R_1^2 + R_2^2)$, where $R_1 = 30.0$ cm and $R_2 = 50.0$ cm. The free-body diagrams for the stone and for the pulley are given in Figure 10.11. Apply $\Sigma \vec{F} = m\vec{a}$ to the stone. The stone accelerates downward, so use coordinates with $+y$ downward. Apply $\Sigma \tau = I\alpha$ to the rotation of the pulley, with clockwise as the positive sense of rotation. n is the normal force applied to the pulley by the axle. $a = R_2\alpha$.

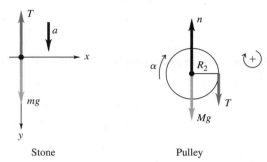

Stone   Pulley

Figure 10.11

Solve: (a) $\Sigma F_y = ma_y$ applied to the stone gives $mg - T = ma$. $\Sigma \tau = I\alpha$ applied to the pulley gives $TR_2 = I\alpha$.

$$I = \tfrac{1}{2}M(R_1^2 + R_2^2) \text{ and } \alpha = a/R_2, \text{ so } T = \tfrac{1}{2}M(1 + [R_1/R_2]^2)a$$

Combining the two equations to eliminate T gives

$$a = \frac{g}{1 + \frac{1}{2}(M/m)(1 + [R_1/R_2]^2)} = \frac{9.80 \text{ m/s}^2}{1 + \frac{1}{2}(10.0 \text{ kg}/2.00 \text{ kg})(1 + [30 \text{ cm}/50 \text{ cm}]^2)}$$

$$a = 2.23 \text{ m/s}^2$$

(b) $T = \frac{1}{2}(10.0 \text{ kg})(1 + [30 \text{ cm}/50 \text{ cm}]^2)(2.23 \text{ m/s}^2) = 15.2$ N

(c) $\alpha = a/R_2 = (2.23 \text{ m/s}^2)/(0.500 \text{ m}) = 4.46$ rad/s^2

Reflect: The tension in the wire is less than the weight of the stone $(19.6$ N$)$. For the stone, the downward force is greater than the upward force and the stone accelerates downward.

10.13. Set Up: For the pulley $I = \frac{1}{2}MR^2$. The elevator has

$$m_1 = \frac{22{,}500 \text{ N}}{9.80 \text{ m/s}^2} = 2300 \text{ kg}.$$

The free-body diagrams for the elevator, the pulley and the counterweight are given in Figure 10.13. Apply $\Sigma \vec{F} = m\vec{a}$ to the elevator and to the counterweight. For the elevator take $+y$ upward and for the counterweight take $+y$ downward, in each case in the direction of the acceleration of the object. Apply $\Sigma \tau = I\alpha$ to the pulley, with clockwise as the positive sense of rotation. n is the normal force applied to the pulley by the axle. The elevator and counterweight each have acceleration a. $a = R\alpha$.

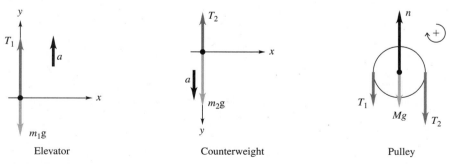

Elevator   Counterweight   Pulley

Figure 10.13

Solve: **(a)** and **(b)** Calculate the acceleration of the elevator: $y - y_0 = v_{0y}t + \frac{1}{2}a_yt^2$ gives

$$a = \frac{2(y - y_0)}{t^2} = \frac{2(6.75 \text{ m})}{(3.00 \text{ s})^2} = 1.50 \text{ m/s}^2$$

$\Sigma F_y = ma_y$ for the elevator gives $T_1 - m_1g = m_1a$ and

$$T_1 = m_1(a + g) = (2300 \text{ kg})(1.50 \text{ m/s}^2 + 9.80 \text{ m/s}^2) = 2.60 \times 10^4 \text{ N}$$

$\Sigma\tau = I\alpha$ for the pulley gives $(T_2 - T_1) = (\frac{1}{2}MR^2)\alpha$. With $\alpha = a/R$ this becomes

$$T_2 - T_1 = \frac{1}{2}Ma. \ T_2 = T_1 + \frac{1}{2}Ma = 2.60 \times 10^4 \text{ N} + \frac{1}{2}(875 \text{ kg})(1.50 \text{ m/s}^2) = 2.67 \times 10^4 \text{ N}$$

$\Sigma F_y = ma_y$ for the counterweight gives $m_2g - T_2 = m_2a$ and

$$m_2 = \frac{T_2}{g - a} = \frac{2.67 \times 10^4 \text{ N}}{9.80 \text{ m/s}^2 - 1.50 \text{ m/s}^2} = 3.22 \times 10^3 \text{ kg}$$

and $w = 3.16 \times 10^4$ N.

Reflect: The tension in the cable must be different on either side of the pulley in order to produce the net torque on the pulley required to give it an angular acceleration. The tension in the cable attached to the elevator is greater than the weight of the elevator and the elevator accelerates upward. The tension in the cable attached to the counterweight is less than the weight of the counterweight and the counterweight accelerates downward.

10.15. Set Up: The free-body diagram for the ball is shown in Figure 10.15. Use the coordinates shown. Take clockwise torques to be positive. For a hollow sphere, $I_{cm} = \frac{2}{3}MR^2$. $a_{cm} = R\alpha$. $R = 0.176$ m.

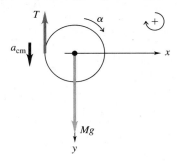

Figure 10.15

Solve: **(a)** Apply $\Sigma F_y = ma_y$ to the motion of the center of mass: $Mg - T = Ma_{cm}$. Apply $\Sigma\tau = I\alpha$ to the rotation about the center of mass: $TR = I_{cm}\alpha$.

$$TR = (\tfrac{2}{3}MR^2)\left(\frac{a_{cm}}{R}\right)$$

and $T = \frac{2}{3}Ma_{cm}$. Combining these two equations to eliminate T gives $Mg - \frac{2}{3}Ma_{cm} = Ma_{cm}$ and $a_{cm} = \frac{3}{5}g = 5.88$ m/s².

(b) $T = \frac{2}{3}Ma_{cm} = \frac{2}{3}(12.5 \text{ kg})(5.88 \text{ m/s}^2) = 49.0 \text{ N}$

(c) $\alpha = \dfrac{a_{cm}}{R} = \dfrac{5.88 \text{ m/s}^2}{0.176 \text{ m}} = 33.4 \text{ rad/s}^2$

Reflect: The tension T is less than the weight $w = Mg = 122.5$ N and the sphere accelerates downward.

10.19. Set Up: **(a)** Let the direction of rotation of the merry-go-round be positive. Apply $\Sigma\tau = I\alpha$ to the merry-go-round to calculate α and then use a constant acceleration equation to calculate ω after 20.0 s. **(b)** $W = \tau\,\Delta\theta$. Calculate $\Delta\theta = \theta - \theta_0$ from a constant acceleration equation. **(c)** $P_{av} = \dfrac{W}{\Delta t}$.

Solve: **(a)** $\Sigma\tau = I\alpha$ gives

$$\alpha = \frac{\Sigma\tau}{I} = \frac{FR}{I} = \frac{(25.0 \text{ N})(4.40 \text{ m})}{24.5 \text{ kg} \cdot \text{m}^2} = 4.49 \text{ rad/s}^2.$$

$$\omega = \omega_0 + \alpha t = 0 + (4.49 \text{ rad/s}^2)(20.0 \text{ s}) = 89.8 \text{ rad/s}$$

(b) $\theta - \theta_0 = \omega_0 t + \frac{1}{2}\alpha t^2 = \frac{1}{2}(4.49 \text{ rad/s}^2)(20.0 \text{ s})^2 = 898 \text{ rad}$

$$W = \tau \, \Delta\theta = (25.0 \text{ N})(4.40 \text{ m})(898 \text{ rad}) = 9.88 \times 10^4 \text{ J}$$

(c) $P_{av} = \dfrac{W}{\Delta t} = \dfrac{9.88 \times 10^4 \text{ J}}{20.0 \text{ s}} = 4.94 \times 10^3 \text{ W}$

Reflect: The power applied when $\omega = 89.8 \text{ rad/s}$, at the end of the 20.0 s time interval, is

$$P = \tau\omega = (25.0 \text{ N})(4.40 \text{ m})(89.8 \text{ rad/s}) = 9.88 \times 10^3 \text{ W}.$$

At $t = 0$, $\omega = 0$ and $P = 0$. ω and hence P increase linearly in time, so

$$P_{av} = \frac{P_f - P_i}{2} = \frac{9.88 \times 10^3 \text{ W}}{2} = 4.94 \times 10^3 \text{ W}.$$

This agrees with our answer in part (c).

10.21. Set Up: $P = \tau\omega$, with ω in rad/s. $\omega = 4000.0 \text{ rev/min} = 419 \text{ rad/s}$. The speed v of the weight is $v = R\omega$. Since ω and v are constant, the tension in the rope equals the weight of the object.

Solve: (a) $\tau = \dfrac{P}{\omega} = \dfrac{150 \times 10^3 \text{ W}}{419 \text{ rad/s}} = 358 \text{ N} \cdot \text{m}$

(b) $\tau = wR$ so $w = \dfrac{\tau}{R} = \dfrac{358 \text{ N} \cdot \text{m}}{0.200 \text{ m}} = 1790 \text{ N}$

(c) $v = R\omega = (0.200 \text{ m})(419 \text{ rad/s}) = 83.8 \text{ m/s}$

Reflect: The rate at which the weight gains gravitational potential energy, wv, equals the power output of the motor, so

$$v = \frac{P}{w} = \frac{150 \times 10^3 \text{ W}}{1790 \text{ N}} = 83.8 \text{ m/s},$$

which agrees with the result we calculated for part (c).

10.25. Set Up: $l = 1.6 \text{ m}$

Solve: (a) $v = 0$ so $L = 0$

(b) Find v: $v_y^2 = v_{0y}^2 + 2a_y(y - y_0)$ gives $v = \sqrt{2g(y - y_0)} = \sqrt{2(9.80 \text{ m/s}^2)(2.5 \text{ m})} = 7.0 \text{ m/s}$

$$L = mvl = (4.0 \text{ kg})(7.0 \text{ m/s})(1.6 \text{ m}) = 44.8 \text{ kg} \cdot \text{m}^2/\text{s}, \text{ counterclockwise}$$

Reflect: The angular momentum increases as the brick falls because of the external torque of the gravity force.

10.27. Set Up: Figure 10.27 shows the force on the drawbridge. The weight acts at the center of the drawbridge and this force has moment arm l. Take clockwise rotation to be positive.

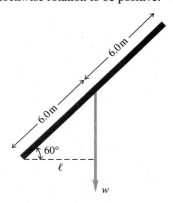

Figure 10.27

Solve: (a) $\tau = wl = (15,000 \text{ N})(6.0 \text{ m})(\cos 60.0°) = 45,000 \text{ N} \cdot \text{m}$

(b) $\dfrac{\Delta L}{\Delta t} = \Sigma\tau = 45,000 \text{ kg} \cdot \text{m}^2/\text{s}^2$

Reflect: The torque changes as the angle of the bridge above the horizontal changes.

10.31. Set Up: The system before and after the collision is sketched in Figure 10.31. The gate has $I = \frac{1}{3}ML^2$. Take counterclockwise torques to be positive.

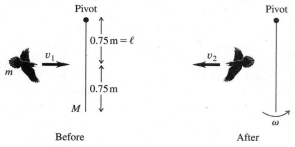

Before   After

Figure 10.31

Solve: (a) The gravity forces exert no torque at the moment of collision and angular momentum is conserved. $L_1 = L_2$. $mv_1l = -mv_2l + I_{gate}\omega$ with $l = L/2$.

$$\omega = \frac{m(v_1 + v_2)l}{\frac{1}{3}ML^2} = \frac{3m(v_1 + v_2)}{2ML} = \frac{3(1.1\text{ kg})(5.0\text{ m/s} + 2.0\text{ m/s})}{2(4.5\text{ kg})(1.5\text{ m})} = 1.71\text{ rad/s}$$

(b) Linear momentum is not conserved; there is an external force exerted by the pivot. But the force on the pivot has zero torque. There is no external torque and angular momentum is conserved.

Reflect: $K_1 = \frac{1}{2}(1.1\text{ kg})(5.0\text{ m/s})^2 = 13.8\text{ J}$.

$$K_2 = \frac{1}{2}(1.1\text{ kg})(2.0\text{ m/s})^2 + \frac{1}{2}(\frac{1}{3}[4.5\text{ kg}][1.5\text{ m/s}]^2)(1.71\text{ rad/s})^2 = 7.1\text{ J}$$

This is an inelastic collision and $K_2 < K_1$.

10.33. Set Up: $L_i = L_f$. $I_{child} = mr^2$, where r is his distance from the axis. $T_i = 6.00$ s so

$$\omega_i = \frac{2\pi\text{ rad}}{6.00\text{ s}} = 1.047\text{ rad/s}.$$

Solve: $I_i = 1200\text{ kg}\cdot\text{m}^2$, since the child is at $r = 0$.

$$I_f = 1200\text{ kg}\cdot\text{m}^2 + (40.0\text{ kg})(2.00\text{ m})^2 = 1360\text{ kg}\cdot\text{m}^2$$

$I_i\omega_i = I_f\omega_f$ gives

$$\omega_f = \left(\frac{I_i}{I_f}\right)\omega_i = \left(\frac{1200\text{ kg}\cdot\text{m}^2}{1360\text{ kg}\cdot\text{m}^2}\right)(1.047\text{ rad/s}) = 0.924\text{ rad/s}.$$

Reflect: When some of the mass moves farther from the axis, the moment of inertia increases and the angular velocity decreases.

10.39. Set Up: The free-body diagram is given in Figure 10.39. F_f is the force on each foot and F_h is the force on each hand. Use coordinates as shown. Take the pivot at his feet and let counterclockwise torques to be positive.

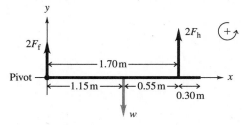

Figure 10.39

Solve: $\Sigma \tau = 0$ gives $+(2F_h)(1.70 \text{ m}) - w(1.15 \text{ m}) = 0$

$$F_h = w\frac{1.15 \text{ m}}{2(1.70 \text{ m})} = 0.338w = 272 \text{ N}$$

$\Sigma F_y = 0$ gives $2F_f + 2F_h - w = 0$ and $F_f = \frac{1}{2}w - F_h = 402 \text{ N} - 272 \text{ N} = 130 \text{ N}$
Reflect: His center of mass is closer to his hands than to his feet, so his hands exert a greater force.

10.43. Set Up: The free-body diagram for the beam is given in Figure 11.43. H_v and H_h are the vertical and horizontal components of the force exerted on the beam at the wall (by the hinge). Since the beam is uniform, its center of gravity is 2.00 m from each end. The angle θ has $\cos\theta = 0.800$ and $\sin\theta = 0.600$. The tension T has been replaced by its x and y components.

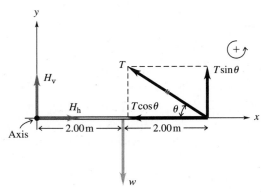

Figure 10.43

Solve: (a) H_v, H_h and $T_x = T\cos\theta$ all produce zero torque. $\Sigma\tau = 0$ gives $-w(2.00 \text{ m}) - w_{\text{load}}(4.00 \text{ m}) + T\sin\theta(4.00 \text{ m}) = 0$ and

$$T = \frac{(150 \text{ N})(2.00 \text{ m}) + (300 \text{ N})(4.00 \text{ m})}{(4.00 \text{ m})(0.600)} = 625 \text{ N}.$$

(b) $\Sigma F_x = 0$ gives $H_h = T\cos\theta = 0$ and $H_h = (625 \text{ N})(0.800) = 500 \text{ N}$. $\Sigma F_y = 0$ gives $H_v - w - w_{\text{load}} + T\sin\theta = 0$ and

$$H_v = w + w_{\text{load}} - T\sin\theta = 150 \text{ N} + 300 \text{ N} - (625 \text{ N})(0.600) = 75 \text{ N}.$$

Reflect: For an axis at the right-hand end of the beam, only w and H_v produce torque. The torque due to w is counterclockwise so the torque due to H_v must be clockwise. To produce a counterclockwise torque, H_v must be upward, in agreement with our result from $\Sigma F_y = 0$.

10.45. Set Up: There is a normal force n_w at the wall. At the floor there is a normal force n and a friction force f. Let the length of the ladder be L. The center of gravity of the ladder is a distance $L/2$ from each end. Let $+y$ be upward and let $+x$ be horizontal, toward the wall. Let the pivot be at the foot of the ladder and let counterclockwise torques be positive.
Solve: (a) The free-body diagram is given in Figure 10.45.

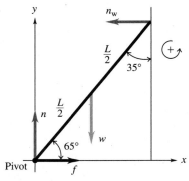

Figure 10.45

(b) $\Sigma F_y = 0$ gives $n - w = 0$ and $n = 250$ N. $\Sigma \tau = 0$ gives

$$n_wL\cos 35° - w\left(\frac{L}{2}\sin 35°\right) = 0$$

$$n_w = \tfrac{1}{2}w\frac{\sin 35°}{\cos 35°} = \tfrac{1}{2}w\tan 35° = 87.5 \text{ N}$$

(c) $\Sigma F_x = 0$ gives $f - n_w = 0$ and $f = n_w = 87.5$ N
Reflect: The friction force required increases when the angle between the ladder and the wall increases, when the ladder becomes closer to horizontal.

10.47. Set Up: The distance along the beam from the hinge to where the cable is attached is 3.0 m. The angle ϕ that the cable makes with the beam is given by

$$\sin\phi = \frac{4.0 \text{ m}}{5.0 \text{ m}}$$

so $\phi = 53.1°$. The center of gravity of the beam is 4.5 m from the hinge. Use coordinates with $+y$ upward and $+x$ to the right. Take the pivot at the hinge and let counterclockwise torque be positive. Express the hinge force as components H_v and H_h. Assume H_v is downward and that H_h is to the right. If one of these components is actually in the opposite direction we will get a negative value for it. Set the tension in the cable equal to its maximum possible value, $T = 1.00$ kN.
Solve: **(a)** The free-body diagram is given in Figure 10.47. T has been resolved into its x and y components.

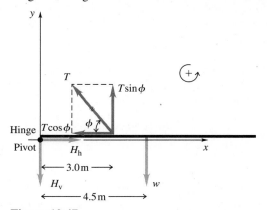

Figure 10.47

(b) $\Sigma \tau = 0$ gives $(T\sin\phi)(3.0 \text{ m}) - w(4.5 \text{ m}) = 0$

$$w = \frac{(T\sin\phi)(3.0 \text{ m})}{4.5 \text{ m}} = \frac{(1.00 \text{ kN})(\sin 53.1°)(3.0 \text{ m})}{4.5 \text{ m}} = 530 \text{ N}$$

(c) $\Sigma F_x = 0$ gives $H_h - T\cos\phi = 0$ and $H_h = (1.00 \text{ kN})(\cos 53.1°) = 600$ N
$\Sigma F_y = 0$ gives $T\sin\phi - H_v - w = 0$ and $H_v = (1.00 \text{ kN})(\sin 53.1°) - 530 \text{ N} = 270$ N
Reflect: $T\cos\phi$, H_v and H_h all have zero moment arms for a pivot at the hinge and therefore produce zero torque. If we consider a pivot at the point where the cable is attached we can see that H_v must be downward to produce a torque that opposes the torque due to w.

10.51. Set Up: The free-body diagram for the door is given in Figure 10.51. U_v, U_h and L_v, L_h are the vertical and horizontal components of the forces exerted on the door by the upper and lower hinges, respectively. Since each hinge supports half the total weight of the door, $U_v = L_v = w/2 = 140$ N.

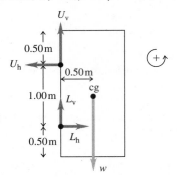

Figure 10.51

Solve: $\Sigma\tau = 0$ with the axis at the lower hinge gives $-w(0.50 \text{ m}) + U_h(1.00 \text{ m}) = 0$ and

$$U_h = \left(\frac{0.50 \text{ m}}{1.00 \text{ m}}\right)(280 \text{ N}) = 140 \text{ N}.$$

$\Sigma F_x = 0$ says $L_h = U_h = 140$ N. The upper hinge exerts a horizontal force of 140 N away from the door and the lower hinge exerts a horizontal force of 140 N toward the door.

Reflect: Our calculation and Newton's third law shows that the door pulls outward on the upper hinge; our experience with doors agrees with this.

10.53. Set Up: Take the pivot at the shoulder joint and let counterclockwise torques be positive. Let $+y$ be upward and let $+x$ be to the right. w_u and w_f are the weights of his upper arm and forearm, respectively. $w = (2.50 \text{ kg})(9.80 \text{ m/s}^2) = 24.5$ N is the weight held in his hand.

Solve: (a) The free-body diagram is given in Figure 10.53.

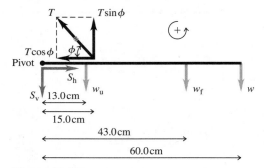

Figure 10.53

(b) $\Sigma\tau = 0$ gives $T\sin\phi(15.0 \text{ cm}) - w_u(13.0 \text{ cm}) - w_f(43.0 \text{ cm}) - w(60.0 \text{ cm}) = 0$

$$T = \frac{[(2.63 \text{ kg})(13.0 \text{ cm}) + (2.44 \text{ kg})(43.0 \text{ cm}) + (2.50 \text{ kg})(60.0 \text{ cm})][9.80 \text{ m/s}^2]}{(15.0 \text{ cm})\sin 14.0°} = 781 \text{ N}$$

Reflect: The force T from the muscle has a small perpendicular component and a small level arm, so its magnitude is much larger than the total weight being supported.

10.57. Set Up: $I = MR^2$, with $R = 2.5 \times 10^{-2}$ m. 1.0×10^{-6} degree $= 1.745 \times 10^{-8}$ rad. $\omega = 19,200$ rpm $= 2.01 \times 10^3$ rad/s. $t = 5.0$ h $= 1.8 \times 10^4$ s.

Solve: $\Omega = \dfrac{\Delta\phi}{\Delta t} = \dfrac{1.745 \times 10^{-8} \text{ rad}}{1.8 \times 10^4 \text{ s}} = 9.694 \times 10^{-13}$ rad/s

$\Omega = \dfrac{\tau}{I\omega}$ so $\tau = \Omega I\omega = \Omega MR^2\omega$

$$\tau = (9.694 \times 10^{-13} \text{ rad/s})(2.0 \text{ kg})(2.5 \times 10^{-2} \text{ m})^2(2.01 \times 10^3 \text{ rad/s}) = 2.4 \times 10^{-12} \text{ N} \cdot \text{m}$$

Reflect: The external torque must be very small for this degree of stability.

10.61. Set Up: Apply $\Sigma\tau = 0$ to the roller, with counterclockwise torque positive and the axis at the point of contact between the roller and the brick. When the roller is on the verge of rolling over the brick, the normal force on it from the ground becomes zero.

Solve: (a) The free-body diagram for the roller is given in Figure 10.61a. B_h and B_v are the componets of the force exerted by the brick.

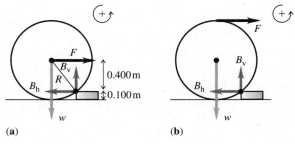

Figure 10.61

The moment arm for w is $\sqrt{(0.500 \text{ m})^2 - (0.400 \text{ m})^2} = 0.300$ m.
The moment arm for F is 0.400 m. $\Sigma\tau = 0$ gives $w(0.300 \text{ m}) - F(0.400 \text{ m}) = 0$ and

$$F = \left(\frac{0.300 \text{ m}}{0.400 \text{ m}}\right)(520 \text{ N}) = 390 \text{ N}.$$

(b) The free-body diagram for the roller is given in Figure 10.61b. Now the moment arm for F is $0.500 \text{ m} + 0.400 \text{ m} = 0.900$ m. $\Sigma\tau = 0$ gives $w(0.300 \text{ m}) - F(0.900 \text{ m}) = 0$ and

$$F = \left(\frac{0.300 \text{ m}}{0.900 \text{ m}}\right)(520 \text{ N}) = 173 \text{ N}.$$

Reflect: Less force is required when the force is applied at the top of the roller.

10.63. Set Up: The weight on the front wheels is n_f, the normal force exerted by the ground on the front wheels. The weight on the rear wheels is n_r, the normal force exerted by the ground on the rear wheels. When the front wheels come off the ground, $n_f \rightarrow 0$. The free-body diagram for the truck without the box is given in Figure 11.63a and with the box in Figure 11.63b. The center of gravity of the truck, without the box is a distance x from the rear wheels.

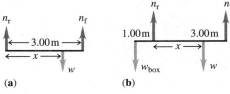

Figure 10.63

Solve: $\sum F_y = 0$ in Figure 11.63a gives $w = n_r + n_f = 8820 \text{ N} + 10{,}780 \text{ N} = 19{,}600 \text{ N}$

$\sum \tau = 0$ in Figure 11.63a, with the axis at the rear wheels and counterclockwise torques positive, gives $n_f(3.00 \text{ m}) - wx = 0$ and

$$x = \frac{n_f(3.00 \text{ m})}{w} = \left(\frac{10{,}780 \text{ N}}{19{,}600 \text{ N}}\right)(3.00 \text{ m}) = 1.65 \text{ m}.$$

(a) $\sum \tau = 0$ in Figure 11.63b, with the axis at the rear wheels and counterclockwise torques positive, gives $w_{\text{box}}(1.00 \text{ m}) + n_f(3.00 \text{ m}) - w(1.65 \text{ m}) = 0.$

$$n_f = \frac{-(3600 \text{ N})(1.00 \text{ m}) + (19{,}600 \text{ N})(1.65 \text{ m})}{3.00 \text{ m}} = 9{,}580 \text{ N}$$

$\sum F_y = 0$ gives $n_r + n_f = w_{\text{box}} + w$ and $n_r = 3600 \text{ N} + 19{,}600 \text{ N} - 9580 \text{ N} = 13{,}620 \text{ N}$. There is 9,580 N on the front wheels and 13,620 N on the rear wheels.

(b) $n_f \to 0$. $\sum \tau = 0$ gives $w_{\text{box}}(1.00 \text{ m}) - w(1.65 \text{ m}) = 0$ and $w_{\text{box}} = 1.65w = 3.23 \times 10^4 \text{ N}$.

Reflect: Placing the box on the tailgate in part (b) reduces the normal force exerted at the front wheels.

10.65. Set Up: Take $+y$ upward and $y = 0$ at the ground. The center of mass of the vine is 4.00 m from either end. Treat the motion in three parts: (i) Jane swinging to where the vine is vertical. Apply conservation of energy. (ii) The inelastic collision between Jane and Tarzan. Apply conservation of angular momentum. (iii) The motion of the combined object after the collision. Apply conservation of energy. The vine has $I = \frac{1}{3}m_{\text{vine}}l^2$ and Jane has $I = m_{\text{Jane}}l^2$, so the system of Jane plus vine has $I_{\text{tot}} = \left(\frac{1}{3}m_{\text{vine}} + m_{\text{Jane}}\right)l^2$.

Solve: (a) The initial and final positions of Jane and the vine for the first stage of the motion are sketched in Figure 10.65a. The initial height of the center of the vine is $h_{\text{vine,i}} = 6.50 \text{ m}$ and its final height is $h_{\text{vine,f}} = 4.00 \text{ m}$. Conservation of energy gives $U_i + K_i = U_f + K_f$. $K_i = 0$ so

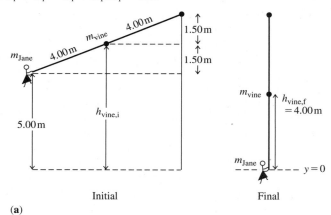

(a)

Figure 10.65a

$$m_{\text{Jane}}g(5.00 \text{ m}) + m_{\text{vine}}g(6.50 \text{ m}) = m_{\text{vine}}g(4.00 \text{ m}) + \tfrac{1}{2}I_{\text{tot}}\omega^2$$

$$\omega = \sqrt{\frac{2[m_{\text{Jane}}(5.00 \text{ m}) + m_{\text{vine}}(2.50 \text{ m})]g}{\left(\frac{1}{3}m_{\text{vine}} + m_{\text{Jane}}\right)l^2}}$$

$$\omega = \sqrt{\frac{2[(60.0 \text{ kg})(5.00 \text{ m}) + (30.0 \text{ kg})(2.50 \text{ m})](9.80 \text{ m/s}^2)}{[\frac{1}{3}(30.0 \text{ kg}) + 60.0 \text{ kg}](8.00 \text{ m})^2}} = 1.28 \text{ rad/s}$$

(b) Conservation of angular momentum applied to the collision gives $L_i = L_f$, so $I_i\omega_i = I_f\omega_f$. $\omega_i = 1.28 \text{ rad/s}$.
$I_i = \left[\frac{1}{3}(30.0 \text{ kg}) + 60.0 \text{ kg}\right](8.00 \text{ m})^2 = 4.48 \times 10^3 \text{ kg} \cdot \text{m}^2$

$$I_f = I_i + m_{\text{Tarzan}}l^2 = 4.48 \times 10^3 \text{ kg} \cdot \text{m}^2 + (72.0 \text{ kg})(8.00 \text{ m})^2 = 9.09 \times 10^3 \text{ kg} \cdot \text{m}^2$$

$$\omega_f = \left(\frac{I_i}{I_f}\right)\omega_i = \left(\frac{4.48 \times 10^3 \text{ kg} \cdot \text{m}^2}{9.09 \times 10^3 \text{ kg} \cdot \text{m}^2}\right)(1.28 \text{ rad/s}) = 0.631 \text{ rad/s}$$

(c) The final position of Tarzan and Jane, when they have swung to their maximum height, is shown in Figure 10.65b. If Tarzan and Jane rise to a height h, then the center of the vine rises to a height $h/2$. Conservation of energy gives $\frac{1}{2}I\omega^2 = (m_{\text{Jane}} + m_{\text{Tarzan}})gh + m_{\text{vine}}gh/2$, where $I = 9.09 \times 10^3 \text{ kg} \cdot \text{m}^2$ and $\omega = 0.631 \text{ rad/s}$, from part (b).

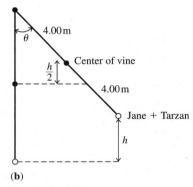

(b)

Figure 10.65b

$$h = \frac{I\omega^2}{2(m_{\text{Jane}} + m_{\text{Tarzan}} + 0.5m_{\text{vine}})g} = \frac{(9.09 \times 10^3 \text{ kg} \cdot \text{m}^2)(0.631 \text{ rad/s})^2}{2(60.0 \text{ kg} + 72.0 \text{ kg} + 15.0 \text{ kg})(9.80 \text{ m/s}^2)} = 1.26 \text{ m}$$

Reflect: Mechanical energy is lost in the inelastic collision.

10.67. Set Up: A accelerates downward, B accelerates upward and the wheel turns clockwise. Apply $\Sigma F_y = ma_y$ to blocks A and B. Let $+y$ be downward for A and y be upward for B. Apply $\Sigma\tau = I\alpha$ to the wheel, with the clockwise sense of rotation positive. Each block has the same magnitude of acceleration, a, and $a = R\alpha$. Call the tension in the cord between C and A T_A and the tension between C and B T_B.
Solve: For A, $\Sigma F_y = ma_y$ gives $m_A g - T_A = m_A a$. For B, $\Sigma F_y = ma_y$ gives $T_B - m_B g = m_B a$. For the wheel,

$T_A R - T_B R = I\alpha = I(a/R)$ and $T_A - T_B = \left(\frac{I}{R^2}\right)a$. Adding these three equations gives

$$(m_A - m_B)g = \left(m_A + m_B + \frac{I}{R^2}\right)a.$$

$$a = \left(\frac{m_A - m_B}{m_A + m_B + I/R^2}\right)g = \left(\frac{4.00 \text{ kg} - 2.00 \text{ kg}}{4.00 \text{ kg} + 2.00 \text{ kg} + (0.300 \text{ kg})/(0.120 \text{ m})^2}\right)(9.80 \text{ m/s}^2) = 0.730 \text{ m/s}^2$$

$$\alpha = \frac{a}{R} = \frac{0.730 \text{ m/s}^2}{0.120 \text{ m}} = 6.08 \text{ rad/s}^2$$

$$T_A = m_A(g - a) = (4.00 \text{ kg})(9.80 \text{ m/s}^2 - 0.730 \text{ m/s}^2) = 36.3 \text{ N}$$

$$T_B = m_B(g + a) = (2.00 \text{ kg})(9.80 \text{ m/s}^2 + 0.730 \text{ m/s}^2) = 21.1 \text{ N}$$

Reflect: The tensions must be different in order to produce a torque that accelerates the wheel when the blocks accelerate.

10.69. Set Up: The free-body diagram for the drawbridge is given in Figure 10.69. For an axis at the lower end, $I = \frac{1}{3}ml^2$.

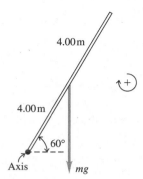

Figure 10.69

Solve: (a) $\Sigma\tau = I\alpha$ gives $mg(4.00 \text{ m})(\cos 60.0°) = \frac{1}{3}ml^2\alpha$ and

$$\alpha = \frac{3g(4.00 \text{ m})(\cos 60.0°)}{(8.00 \text{ m})^2} = 0.919 \text{ rad/s}^2.$$

(b) α depends on the angle the bridge makes with the horizontal. α is not constant during the motion and $\omega = \omega_0 + \alpha t$ cannot be used.

(c) Use conservation of energy. Take $y = 0$ at the lower end of the drawbridge, so $y_i = (4.00 \text{ m})(\sin 60.0°)$ and $y_f = 0$. $K_f + U_f = K_i + U_i + W_{\text{other}}$ gives $U_i = K_f$, $mgy_i = \frac{1}{2}I\omega^2$. $mgy_i = \frac{1}{2}(\frac{1}{3}ml^2)\omega^2$ and

$$\omega = \frac{\sqrt{6gy_i}}{l} = \frac{\sqrt{6(9.80 \text{ m/s}^2)(4.00 \text{ m})(\sin 60.0°)}}{8.00 \text{ m}} = 1.78 \text{ rad/s}.$$

Reflect: If we incorrectly assume that α is constant and has the value calculated in part (a), then $\omega^2 = \omega_0^2 + 2\alpha(\theta - \theta_0)$ gives $\omega = 139 \text{ rad/s}$. The angular acceleration increases as the bridge rotates and the actual angular velocity is larger than this.

10.75. Set Up: (a) The pulley has moment of inertia $I_p = \frac{1}{2}m_pR_p^2$, with $m_p = 2.00 \text{ kg}$ and $R_p = 0.200 \text{ m}$. The cylinder has moment of inertia $I_c = \frac{1}{2}m_cR_c^2$, with $m_c = 5.00 \text{ kg}$ and $R_c = 0.400 \text{ m}$. Apply conservation of energy. Take $+y$ downward and let $y = 0$ at the final position of the bucket. $a = R_p\alpha_p$ and $a = R_c\alpha_c$. **(b)** and **(c)** The free-body-diagram for the bucket is given in Figure 10.75.

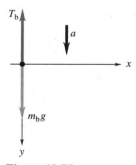

Figure 10.75

Solve: (a) Conservation of energy gives $U_i + K_i = U_f + K_f$. For the bucket, $y_i = 2.00 \text{ m}$ and $y_f = 0$. $K_i = 0$ and $U_f = 0$ so the conservation of energy expression becomes $U_i = K_f$. This gives $m_bgy_i = \frac{1}{2}I_p\omega_p^2 + \frac{1}{2}I_c\omega_c^2 + \frac{1}{2}m_bv^2$. We can rewrite the rotational kinetic energies in terms of a:

$$\frac{1}{2}I_p\omega_p^2 + \frac{1}{2}I_c\omega_c^2 = \frac{1}{2}(\frac{1}{2}m_pR_p^2)(a/R_p)^2 + \frac{1}{2}(\frac{1}{2}m_cR_c^2)(a/R_c)^2 = \frac{1}{4}(m_p + m_c)a^2$$

Then $m_b g y_i = \frac{1}{4}(2m_b + m_p + m_c)v^2$ and

$$v = \sqrt{\frac{4m_b g y_i}{2m_b + m_p + m_c}} = \sqrt{\frac{4(3.00\ \text{kg})(9.80\ \text{m/s}^2)(2.00\ \text{m})}{6.00\ \text{kg} + 2.00\ \text{kg} + 5.00\ \text{kg}}} = 4.25\ \text{m/s}$$

(b) $v_y^2 = v_{0y}^2 + 2a_y(y - y_0)$ gives

$$a_y = \frac{v_y^2 - v_{0y}^2}{2(y - y_0)} = \frac{(4.25\ \text{m/s})^2 - 0}{2(2.00\ \text{m})} = 4.52\ \text{m/s}^2$$

(c) Let T_b be the tension in the segment of cable between the pulley and the bucket and let T_c be the tension in the segment of cable between the pulley and the cylinder. $\Sigma F_y = ma_y$ for the bucket gives $m_b g - T_b = m_b a$ and

$$T_b = m_b(g - a) = (3.00\ \text{kg})(9.80\ \text{m/s}^2 - 4.52\ \text{m/s}^2) = 15.8\ \text{N}$$

$\Sigma\tau = I\alpha$ for the pulley gives $(T_b - T_c)R_p = \frac{1}{2}m_p R_p^2(a/R_p)$. $T_b - T_c = \frac{1}{2}m_p a$ and

$$T_c = T_b - \tfrac{1}{2}m_p a = 15.8\ \text{N} - \tfrac{1}{2}(2.00\ \text{kg})(4.52\ \text{m/s}^2) = 11.3\ \text{N}.$$

Or, $\Sigma\tau = I\alpha$ for the cylinder gives $T_c R_c = \frac{1}{2}m_c R_c^2(a/R_c)$ and

$$T_c = \tfrac{1}{2}m_c a = \tfrac{1}{2}(5.00\ \text{kg})(4.52\ \text{m/s}^2) = 11.3\ \text{N},$$

which checks.

Reflect: The acceleration of the bucket is less than g because the weight of the bucket must accelerate not only the bucket but also the pulley and the cylinder.

10.79. Set Up: The angular momentum of Sedna is conserved as it moves in its orbit.
Solve: (a) $L = mvl$ so $v_1 l_1 = v_2 l_2$. When $v_1 = 4.64\ \text{km/s}$, $l_1 = 76\ \text{AU}$.

$$v_2 = v_1\left(\frac{l_1}{l_2}\right) = (4.64\ \text{km/s})\left(\frac{76\ \text{AU}}{942\ \text{AU}}\right) = 0.374\ \text{km/s}$$

(b) Since vl is constant the maximum speed is at the minimum distance and the minimum speed is at the maximum distance.

(c) $\dfrac{K_1}{K_2} = \dfrac{\frac{1}{2}mv_1^2}{\frac{1}{2}mv_2^2} = \left(\dfrac{v_1}{v_2}\right)^2 = \left(\dfrac{l_2}{l_1}\right)^2 = \left(\dfrac{942\ \text{AU}}{76\ \text{AU}}\right)^2 = 154$

Reflect: Since the units of l cancel in the ratios there is no need to convert from AU to m. The gravity force of the sun does work on Sedna as it moves toward or away from the sun and this changes the kinetic energy during the orbit. But this force exerts no torque, so the angular momentum of Sedna is constant.

10.81. Set Up: The thermal energy developed equals the decrease in kinetic energy of the system. The angular momentum of the system is conserved. $I_B = 3I_A$.
Solve: $L_i = L_f$ gives $I_A \omega_0 = (I_A + I_B)\omega_f$ and

$$\omega_f = \left(\frac{I_A}{I_A + I_B}\right)\omega_0 = \frac{I_A}{4I_A}\omega_0 = \omega_0/4.$$

$$K_i = \tfrac{1}{2}I_A\omega_0^2.$$

$$K_f = \tfrac{1}{2}(I_A + I_B)\omega_f^2 = \tfrac{1}{2}(4I_A)\left(\frac{\omega_0}{4}\right)^2 = \tfrac{1}{4}K_i.$$

$K_i - K_f = \frac{3}{4}K_i = 2400\ \text{J}$ and $K_i = 3200\ \text{J}$.
Reflect: Angular momentum is conserved because there is no external torque due to the force that pushes the shafts together, but the friction forces in the clutch mechanism do negative work and remove kinetic energy.

10.83. Set Up: Figure 10.83a shows the distances and angles. $\theta + \phi = 90°$. $\theta = 56.3°$ and $\phi = 33.7°$. The distances x_1 and x_2 are $x_1 = (90 \text{ cm})\cos\theta = 50.0$ cm and $x_2 = (135 \text{ cm})\cos\phi = 112$ cm.

The free-body diagram for the person is given in Figure 10.83b. $w_l = 277$ N is the weight of his feet and legs and $w_t = 473$ N is the weight of his trunk. n_f and f_f are the total normal and friction forces exerted on his feet and n_h and f_h are those forces on his hands.

The free-body diagram for his legs is given in Figure 10.83c. F is the force exerted on his legs by his hip joints.

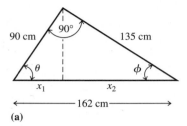

(a)

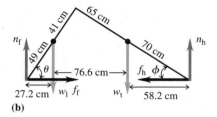

(b)

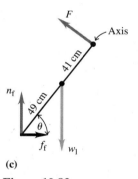

(c)

Figure 10.83

Solve: (a) Consider the force diagram of Figure 10.83b. $\Sigma\tau = 0$ with the pivot at his feet and counterclockwise torques positive gives

$$n_h(162 \text{ cm}) - (277 \text{ N})(27.2 \text{ cm}) - (473 \text{ N})(103.8 \text{ cm}) = 0$$

$n_h = 350$ N, so there is a normal force of 175 N at each hand. $n_f + n_h - w_l - w_t = 0$ so $n_f = w_l + w_t - n_h = 750 \text{ N} - 350 \text{ N} = 400$ N, so there is a normal force of 200 N at each foot.
(b) Consider the force diagram of Figure 10.83c. $\Sigma\tau = 0$ with the pivot at his hips and counterclockwise torques positive gives $f_f(74.9 \text{ cm}) + w_l(22.8 \text{ cm}) - n_f(50.0 \text{ cm}) = 0$

$$f_f = \frac{(400 \text{ N})(50.0 \text{ cm}) - (277 \text{ N})(22.8 \text{ cm})}{74.9 \text{ cm}} = 183 \text{ N}$$

There is a friction force of 92 N at each foot. $\Sigma F_x = 0$ in Figure 10.83b gives $f_h = f_f$, so there is a friction force of 92 N at each hand.
Reflect: In this position the normal forces at his feet and at his hands don't differ very much.

ELASTICITY AND PERIODIC MOTION

Problems 1, 7, 9, 11, 15, 19, 21, 23, 29, 31, 33, 37, 41, 43, 45, 47, 51, 55, 57, 59, 61, 63, 69, 71

Solutions to Problems

11.1. Set Up: $A = \pi r^2$, with $r = 2.75 \times 10^{-4}$ m. The force F_T applied to the end of the wire is $mg = (25.0 \text{ kg})(9.80 \text{ m/s}^2) = 245$ N. stress $= F_T/A$ and strain $= \Delta l/l_0$. $Y = $ stress/strain

Solve: (a) stress $= \dfrac{245 \text{ N}}{\pi(2.75 \times 10^{-4} \text{ m})^2} = 1.03 \times 10^9$ Pa

(b) strain $= \Delta l/l_0 = \dfrac{1.10 \times 10^{-3} \text{ m}}{0.750 \text{ m}} = 1.47 \times 10^{-3}$

(c) $Y = \dfrac{\text{stress}}{\text{strain}} = \dfrac{1.03 \times 10^9 \text{ Pa}}{1.47 \times 10^{-3}} = 7.01 \times 10^{11}$ Pa

Reflect: Our result for Y is about the same size as the values in Table 11.1. In SI units the stress is very large. The strain is dimensionless and small, so Y is very large and has the same units (Pa) as stress.

11.7. Set Up: For steel, $Y = 2.0 \times 10^{11}$ Pa. $A = \pi d^2/4$.

Solve: $Y = \dfrac{l_0 F_\perp}{A \, \Delta l}$ gives $A = \dfrac{l_0 F_\perp}{Y \Delta l} = \dfrac{(2.00 \text{ m})(400.0 \text{ N})}{(2.0 \times 10^{11} \text{ Pa})(0.25 \times 10^{-2} \text{ m})} = 1.6 \times 10^{-6} \text{ m}^2$

$$d = \sqrt{\frac{4A}{\pi}} = \sqrt{\frac{4(1.6 \times 10^{-6} \text{ m}^2)}{\pi}} = 1.43 \text{ mm}$$

Reflect: The thinner the wire the more it stretches for the same applied force. The length of this wire changes by 0.12%

11.9. Set Up: $Y = \dfrac{F_T/A}{\Delta l/l_0}$ so $F_T = \left(\dfrac{YA}{l_0}\right)\Delta l$ and $k = \dfrac{YA}{l_0}$. From Problem 11.8, $k = 4.6 \times 10^5$ N/m for the natural Achilles tendon. $A = \pi r^2$

Solve: (a) $k = \dfrac{YA}{l_0}$ so $A = \dfrac{kl_0}{Y} = \dfrac{(4.6 \times 10^5 \text{ N/m})(0.25 \text{ m})}{30 \times 10^9 \text{ Pa}} = 3.8 \times 10^{-6} \text{ m}^2$

$A = \pi r^2$ so $r = \sqrt{A/\pi} = 1.1$ mm and the diameter is 2.2 mm.

(b) The natural tendon has $r = \sqrt{(78.1 \text{ mm}^2)/\pi} = 4.99$ mm and diameter 10.0 mm. The artificial tendon's diameter is much smaller.

Reflect: The artificial tendon has a larger Y and therefore a smaller diameter.

11.11. Set Up: $= \dfrac{F_T/A}{\Delta l/l_0}$

Solve: (a) $F_T = 8mg = 5880$ N. $\dfrac{\Delta l}{l_0} = \dfrac{F_T}{AY} = \dfrac{5880 \text{ N}}{(10 \times 10^{-4} \text{ m}^2)(24 \times 10^6 \text{ Pa})} = 0.24 = 24\%$

(b) $F_T = 4mg$ so $\dfrac{\Delta l}{l_0} = \frac{1}{2}(24\%) = 12\%$

Reflect: Young's modulus for cartilage is much smaller than typical values for metals and the fractional change in length is larger.

11.15. Set Up: The surface pressure on earth is about 1 atm. $\dfrac{\Delta V}{V_0} = -\dfrac{\Delta p}{B}$

Solve: $\Delta V = -V_0\dfrac{\Delta p}{B} = -(5.00 \text{ L})\dfrac{(91)(1.01 \times 10^5 \text{ Pa})}{5.0 \times 10^9 \text{ Pa}} = -9.1 \text{ mL}$

The volume would decrease 9.1 mL.

Reflect: The percent change in the volume is $\dfrac{\Delta V}{V_0} = 0.18\%$. A nearly 100-fold increase in pressure has a small effect on the volume because the bulk modulus is very large.

11.19. Set Up: $S = \dfrac{F_{\|}}{A\phi}$. $F_{\|} = F\sin 12°$. ϕ is in radians. $F = 8mg$, with $m = 110$ kg. π rad $= 180°$

Solve: $\phi = \dfrac{F_{\|}}{AS} = \dfrac{8mg \sin 12°}{(10 \times 10^{-4} \text{ m}^2)(12 \times 10^6 \text{ Pa})} = 0.1494 \text{ rad} = 8.6°$

Reflect: The shear modulus of cartilage is much less than the values for metals given in Table 11.1.

11.21. Set Up: $\dfrac{F_\perp}{A} = \frac{1}{3}(2.40 \times 10^8 \text{ Pa}) = 0.80 \times 10^8 \text{ Pa}$. The free-body diagram for the elevator is given in Figure 11.21. $F_\perp$ is the tension in the cable.

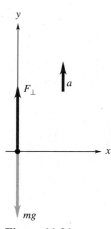

Figure 11.21

Solve: $F_\perp = A(0.80 \times 10^8 \text{ Pa}) = (3.00 \times 10^{-4} \text{ m}^2)(0.80 \times 10^8 \text{ Pa}) = 2.40 \times 10^4 \text{ N}$. $\Sigma F_y = ma_y$ applied to the elevator gives $F_\perp - mg = ma$ and

$$a = \dfrac{F_\perp}{m} - g = \dfrac{2.40 \times 10^4 \text{ N}}{1200 \text{ kg}} - 9.80 \text{ m/s}^2 = 10.2 \text{ m/s}^2$$

Reflect: The tension in the cable is about twice the weight of the elevator.

11.23. Set Up: The frequency f in Hz is the number of cycles per second. The angular frequency ω is $\omega = 2\pi f$ and has units of radians. The period T and frequency f are related by $T = \dfrac{1}{f}$.

Solve: **(a)** $T = \dfrac{1}{f} = \dfrac{1}{466 \text{ Hz}} = 2.15 \times 10^{-3} \text{ s}$. $\omega = 2\pi f = 2\pi(466 \text{ Hz}) = 2.93 \times 10^3 \text{ rad/s}$

(b) $f = \dfrac{1}{T} = \dfrac{1}{50.0 \times 10^{-6} \text{ s}} = 2.00 \times 10^4 \text{ Hz}$. $\omega = 2\pi f = 1.26 \times 10^5 \text{ rad/s}$

(c) $f = \dfrac{\omega}{2\pi}$ so f ranges from $\dfrac{2.7 \times 10^{15} \text{ Hz}}{2\pi \text{ rad}} = 4.3 \times 10^{14}$ Hz to $\dfrac{4.7 \times 10^{15} \text{ Hz}}{2\pi \text{ rad}} = 7.5 \times 10^{14}$ Hz

$T = \dfrac{1}{f}$ so T ranges from $\dfrac{1}{7.5 \times 10^{14} \text{ Hz}} = 1.3 \times 10^{-15}$ s to $\dfrac{1}{4.3 \times 10^{14} \text{ Hz}} = 2.3 \times 10^{-15}$ s

(d) $T = \dfrac{1}{f} = \dfrac{1}{5.0 \times 10^{6} \text{ Hz}} = 2.0 \times 10^{-7}$ s and $\omega = 2\pi f = 2\pi (5.0 \times 10^{6} \text{ Hz}) = 3.1 \times 10^{7}$ rad/s

Reflect: Visible light has much higher frequency than either sounds we can hear or ultrasound. Ultrasound is sound with frequencies higher than what the ear can hear. Large f corresponds to small T.

11.29. Set Up: Velocity, position and total energy are related by $E = \frac{1}{2}kA^2 = \frac{1}{2}mv_x^2 + \frac{1}{2}kx^2$. The maximum speed is at $x = 0$.

Solve: (a) $E = \frac{1}{2}mv_x^2 + \frac{1}{2}kx^2 = \frac{1}{2}(0.150 \text{ kg})(0.30 \text{ m/s})^2 + \frac{1}{2}(300.0 \text{ N/m})(0.012 \text{ m})^2$

$$E = 6.75 \times 10^{-3} \text{ J} + 2.16 \times 10^{-2} \text{ J} = 2.84 \times 10^{-2} \text{ J}.$$

(b) $E = \frac{1}{2}kA^2$ so $A = \sqrt{\dfrac{2E}{k}} = \sqrt{\dfrac{2(2.84 \times 10^{-2} \text{ J})}{300.0 \text{ N/m}}} = 0.0138$ m.

(c) For $x = 0$, $\frac{1}{2}mv_{\max}^2 = E$ and $v_{\max} = \sqrt{\dfrac{2E}{m}} = \sqrt{\dfrac{2(2.84 \times 10^{-2} \text{ J})}{0.150 \text{ kg}}} = 0.615$ m/s.

Reflect: When $x = 0.0120$ m the system has both kinetic and elastic potential energy. When $x = \pm A$ the energy is all elastic potential energy and when $x = 0$ the energy is all kinetic energy.

11.31. Set Up: Velocity, position and total energy are related by $E = \frac{1}{2}kA^2 = \frac{1}{2}mv_x^2 + \frac{1}{2}kx^2$. Acceleration and position are related by $-kx = ma_x$. The maximum magnitude of acceleration is at $x = \pm A$.

Solve: (a) $E = \frac{1}{2}mv_x^2 + \frac{1}{2}kx^2 = \frac{1}{2}(2.00 \text{ kg})(-4.00 \text{ m/s})^2 + \frac{1}{2}(315.0 \text{ N/m})(+0.200 \text{ m})^2$. $E = 16.0$ J $+$ 6.3 J $= 22.3$ J. $E = \frac{1}{2}kA^2$ and

$$A = \sqrt{\dfrac{2E}{k}} = \sqrt{\dfrac{2(22.3 \text{ J})}{315 \text{ N/m}}} = 0.376 \text{ m}.$$

(b) $a_{\max} = \dfrac{k}{m}A = \left(\dfrac{315 \text{ N/m}}{2.00 \text{ kg}}\right)(0.376 \text{ m}) = 59.2$ m/s^2

(c) $F_{\max} = ma_{\max} = (2.00 \text{ kg})(59.2 \text{ m/s}^2) = 118$ N. Or, $F_x = -kx$ gives $F_{\max} = kA = (315 \text{ N/m})(0.376 \text{ m}) = 118$ N, which checks.

Reflect: The maximum force and maximum acceleration occur when the displacement is maximum and the velocity is zero.

11.33. Set Up: $K + U = E$, with $E = \frac{1}{2}kA^2$ and $U = \frac{1}{2}kx^2$

Solve: $U = K$ says $2U = E$. This gives $2(\frac{1}{2}kx^2) = \frac{1}{2}kA^2$, so $x = A/\sqrt{2}$.

Reflect: When $x = A/2$ the kinetic energy is three times the elastic potential energy.

11.37. Set Up: From Problem 11.35, $k = 148$ N/m, $m = 2.40$ kg and $A = 3.0$ cm.

Solve: (a) $\frac{1}{2}mv_x^2 + \frac{1}{2}kx^2 = \frac{1}{2}kA^2$ with $x = 0$ gives $v_{\max} = \sqrt{\dfrac{k}{m}}A = \sqrt{\dfrac{148 \text{ N/m}}{2.40 \text{ kg}}}(3.0 \text{ cm}) = 23.6$ cm/s

(b) $-kx = ma_x$ with $x = \pm A$ gives $a = \dfrac{k}{m}A = \left(\dfrac{148 \text{ N/m}}{2.40 \text{ kg}}\right)(3.0 \text{ cm}) = 185$ cm/s^2

(c) $|v_x| = v_{\max}$ at $x = 0$ and $|a_x| = a_{\max}$ at $x = \pm A$

Reflect: At the point in the motion where the speed is a maximum the acceleration is zero and at the point where the magnitude of the acceleration is a maximum the speed is zero.

11.41. Set Up: The period is the time for 1 cycle; after time T the motion repeats. The graph shows $v_{max} = 20.0$ cm/s. $\frac{1}{2}mv_x^2 + \frac{1}{2}kx^2 = \frac{1}{2}kA^2$

Solve: (a) $T = 1.60$ s

(b) $f = \dfrac{1}{T} = 0.625$ Hz

(c) $\omega = 2\pi f = 3.93$ rad/s

(d) $v_x = v_{max}$ when $x = 0$ so $\frac{1}{2}kA^2 = \frac{1}{2}mv_{max}^2$. $A = v_{max}\sqrt{\dfrac{m}{k}}$. $f = \dfrac{1}{2\pi}\sqrt{\dfrac{k}{m}}$ so $A = v_{max}/(2\pi f)$. From the graph in the problem, $v_{max} = 0.20$ m/s.

$$A = (0.20 \text{ m/s})/(2\pi)(0.625 \text{ Hz}) = 0.0.051 \text{ m} = 5.1 \text{ cm}$$

The mass is at $x = \pm A$ when $v_x = 0$, and this occurs at $t = 0.4$ s, 1.2 s, and 1.8 s.

(e) $-kx = ma_x$ gives

$$a_{max} = \frac{kA}{m} = (2\pi f)^2 A = (4\pi^2)(0.625 \text{ Hz})^2(0.051 \text{ m}) = 0.79 \text{ m/s}^2 = 79 \text{ cm/s}^2.$$

The acceleration is maximum when $x = \pm A$ and this occurs at the times given in (d).

(f) $T = 2\pi\sqrt{\dfrac{m}{k}}$ so $m = k\left(\dfrac{T}{2\pi}\right)^2 = (75 \text{ N/m})\left(\dfrac{1.60 \text{ s}}{2\pi}\right)^2 = 4.9$ kg

Reflect: The speed is maximum at $x = 0$, when $a_x = 0$. The magnitude of the acceleration is maximum at $x = \pm A$, where $v_x = 0$.

11.43. Set Up: $f = \dfrac{1}{2\pi}\sqrt{\dfrac{k}{m}}$

Solve: (a) $f_s = \dfrac{1}{2\pi}\sqrt{\dfrac{k}{m_s}}$, $f_{s+v} = \dfrac{1}{2\pi}\sqrt{\dfrac{k}{m_s + m_v}}$

$$\frac{f_{s+v}}{f_s} = \left(\frac{1}{2\pi}\sqrt{\frac{k}{m_s + m_v}}\right)\left(2\pi\sqrt{\frac{m_s}{k}}\right) = \sqrt{\frac{m_s}{m_s + m_v}} = \frac{1}{\sqrt{1 + m_v/m_s}}$$

(b) $\left(\dfrac{f_{s+v}}{f_s}\right)^2 = \dfrac{1}{1 + m_v/m_s}$. Solving for m_v gives

$$m_v = m_s\left(\left[\frac{f_s}{f_{s+v}}\right]^2 - 1\right) = (2.10 \times 10^{-16} \text{ g})\left(\left[\frac{2.00 \times 10^{15} \text{ Hz}}{2.87 \times 10^{14} \text{ Hz}}\right]^2 - 1\right) = 9.99 \times 10^{-15} \text{ g}$$

$m_V = 9.99$ femtograms

Reflect: When the mass increases, the frequency of oscillation increases.

11.45. Set Up: $T = 2\pi\sqrt{\dfrac{L}{g}}$. The motion specified is one-half of a cycle so the time is $t = T/2$.

Solve: $T = 2\pi\sqrt{\dfrac{3.50 \text{ m}}{9.80 \text{ m/s}^2}} = 3.75$ s, so $t = 1.88$ s

Reflect: The period of a simple pendulum does not depend on its amplitude, as long as the amplitude is small, and does not depend on the mass of the bob. The period depends only on the length of the pendulum and the value of g.

11.47. Set Up: On earth, $g_e = 9.80$ m/s^2.

Solve: $T = 2\pi\sqrt{\dfrac{L}{g}}$ so $T^2 g = 4\pi^2 L = $ constant. $T_M^2 g_M = T_e^2 g_e$ and

$$T_M = T_e\sqrt{\frac{g_e}{g_M}} = (1.60 \text{ s})\sqrt{\frac{9.80 \text{ m/s}^2}{3.71 \text{ m/s}^2}} = 2.60 \text{ s}.$$

Reflect: Smaller g gives a larger period.

11.51. Set Up: As shown in Figure 11.51, the height h above the lowest point of the swing is $h = L - L\cos\theta = L(1 - \cos\theta)$.

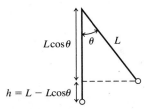

Figure 11.51

Solve: (a) At the maximum angle of swing, $K = 0$ and $E = mgh$.

$$E_i = mgL(1 - \cos\theta_i) = (2.50\ \text{kg})(9.80\ \text{m/s}^2)(1.45\ \text{m})(1 - \cos 11°) = 0.653\ \text{J}$$

$$E_f = mgL(1 - \cos\theta_f) = (2.50\ \text{kg})(9.80\ \text{m/s}^2)(1.45\ \text{m})(1 - \cos 4.5°) = 0.110\ \text{J}$$

The mechanical energy lost is $E_i - E_f = 0.543\ \text{J}$.
(b) The mechanical energy has been converted to other forms by air resistance and by dissipative forces within the rope.
Reflect: After a while the rock will have come to rest and then all its initial mechanical energy has been "lost", has been converted to other forms.

11.55. Set Up: $A = 0.0500\ \text{m}$. $\omega = 3500\ \text{rpm} = 366.5\ \text{rad/s}$. $ma_x = -kx$ so $a_{max} = \dfrac{k}{m}A = \omega^2 A$ is the

magnitude of the acceleration when $x = \pm A$. $v_{max} = \sqrt{\dfrac{k}{m}}A = \omega A$.

Solve: (a) $a_{max} = \omega^2 A = (366.5\ \text{rad/s})^2(0.0500\ \text{m}) = 6.72 \times 10^3\ \text{m/s}^2$
(b) $F_{max} = ma_{max} = (0.450\ \text{kg})(6.72 \times 10^3\ \text{m/s}^2) = 3.02 \times 10^3\ \text{N}$
(c) $v_{max} = \omega A = (366.5\ \text{rad/s})(0.0500\ \text{m}) = 18.3\ \text{m/s}$.

$$K_{max} = \tfrac{1}{2}mv_{max}^2 = \tfrac{1}{2}(0.450\ \text{kg})(18.3\ \text{m/s})^2 = 75.4\ \text{J}$$

Reflect: For a given amplitude, the maximum acceleration and maximum velocity increase when the frequency of the motion increases and the period decreases.

11.57. Set Up: $T = 2\pi\sqrt{\dfrac{L}{g}}$. Relative to the accelerating rocket, the downward acceleration of an object that is

dropped inside the rocket is $a + 9.80\ \text{m/s}^2$; the rock is accelerated downward by gravity and the rocket accelerates upward toward the rock.
Solve: Use the period $T = 2.50\ \text{s}$ when the rocket is at rest to calculate the length of the pendulum.

$$L = g\left(\frac{T}{2\pi}\right)^2 = (9.80\ \text{m/s}^2)\left(\frac{2.50\ \text{s}}{2\pi}\right)^2 = 1.55\ \text{m}$$

When the rocket is accelerating, $T = 2\pi\sqrt{\dfrac{L}{g+a}}$. $g + a = L\left(\dfrac{2\pi}{T}\right)^2 = (1.55\ \text{m})\left(\dfrac{2\pi}{1.25\ \text{s}}\right)^2 = 39.2\ \text{m/s}^2$ and

$a = 39.2\ \text{m/s}^2 - 9.80\ \text{m/s}^2 = 29.4\ \text{m/s}^2$.

Reflect: If the rocket had a downward acceleration, as an elevator might, then the period decreases. If the rocket is in free-fall, with downward acceleration g, then the pendulum bob doesn't swing at all and the period is infinite.

11.59. Set Up: the motion from one position of $v = 0$ to the next is half a period. The period depends only on g and $g = G\dfrac{m}{R^2}$, where m and R are the mass and radius of Newtonia. The circumference c is related to R by $c = 2\pi R$. $G = 6.673 \times 10^{-11} \text{ N} \cdot \text{m}^2/\text{kg}^2$.

Solve: $T = 2(1.42 \text{ s}) = 2.84 \text{ s.}$ $T = 2\pi\sqrt{\dfrac{L}{g}}$ gives

$$g = \left(\frac{2\pi}{T}\right)^2 L = \left(\frac{2\pi}{2.84 \text{ s}}\right)^2 (1.85 \text{ m}) = 9.06 \text{ m/s}^2.$$

$$R = \frac{c}{2\pi} = \frac{5.14 \times 10^7 \text{ m}}{2\pi} = 8.18 \times 10^6 \text{ m.}$$

$$m = \frac{gR^2}{G} = \frac{(9.06 \text{ m/s}^2)(8.18 \times 10^6 \text{ m})^2}{6.673 \times 10^{-11} \text{ N} \cdot \text{m}^2/\text{kg}^2} = 9.08 \times 10^{24} \text{ kg}$$

Reflect: The value of G on Newtonia is within 10% of the value on earth. The mass and radius are somewhat larger than those of earth.

11.61. Set Up: The maximum acceleration of the lower block can't exceed the maximum acceleration that can be given to the other block by the friction force.

Solve: For block m, the maximum friction force is $f_s = \mu_s n = \mu_s mg$. $\Sigma F_x = ma_x$ gives $\mu_s mg = ma$ and $a = \mu_s g$. Then treat both blocks together and consider their simple harmonic motion. $a_{max} = \left(\dfrac{k}{M+m}\right)A$. Set $a_{max} = a$ and solve for A: $\mu_s g = \left(\dfrac{k}{M+m}\right)A$ and $A = \dfrac{\mu_s g(M+m)}{k}$.

Reflect: If A is larger than this the spring gives the block with mass M a larger acceleration than friction can give the other block, and the first block accelerates out from underneath the other block.

11.63. Set Up: For steel, $Y = 2.0 \times 10^{11}$ Pa. $\omega = 2.00 \text{ rev/s} = 12.6 \text{ rad/s}$. The elongation is a small fraction of the length, so use $l_0 = 0.50$ m to calculate a_{rad}.

Solve: Apply $\Sigma F_y = ma_y$ to the mass when it is at the bottom of the circle. The acceleration is $a_{rad} = r\omega^2 = (0.50 \text{ m})(12.6 \text{ rad/s})^2 = 79.4 \text{ m/s}^2$ and is upward, so take $+y$ upward. $F_\perp$ is the tension in the wire. $F_\perp - mg = ma_{rad}$.

$$F_\perp = m(g + a_{rad}) = (15.0 \text{ kg})(9.80 \text{ m/s}^2 + 79.4 \text{ m/s}^2) = 1.34 \times 10^3 \text{ N.}$$

$$Y = \frac{l_0 F_\perp}{A\, \Delta l} \text{ so } \Delta l = \frac{l_0 F_\perp}{AY} = \frac{(0.50 \text{ m})(1.34 \times 10^3 \text{ N})}{(0.010 \times 10^{-4} \text{ m}^2)(2.0 \times 10^{11} \text{ Pa})} = 3.35 \times 10^{-3} \text{ m} = 3.35 \text{ mm.}$$

Reflect: When the mass hangs at rest at the end of the wire, $F_\perp = mg = 147$ N and the wire is stretched 0.37 mm. The radial acceleration is large and $F_\perp$ is much greater than mg.

11.69. Set Up: When the ride is at rest the tension $F_\perp$ in the rod is the weight 1900 N of the car and occupants. When the ride is operating, the tension $F_\perp$ in the rod is obtained by applying $\sum \vec{F} = m\vec{a}$ to a car and its occupants. The free-body diagram is shown in Figure 11.69. The car travels in a circle of radius $r = l\sin\theta$, where l is the length of the rod and θ is the angle the rod makes with the vertical. For steel, $Y = 2.0 \times 10^{11}$ Pa. $\omega = 8.00$ rev/min $= 0.838$ rad/s.

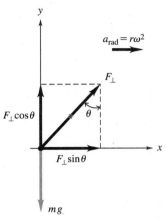

Figure 11.69

Solve: (a) $\Delta l = \dfrac{l_0 F_\perp}{YA} = \dfrac{(15.0\text{ m})(1900\text{ N})}{(2.0 \times 10^{11}\text{ Pa})(8.00 \times 10^{-4}\text{ m}^2)} = 1.78 \times 10^{-4}\text{ m} = 0.18\text{ mm}$

(b) $\sum F_x = ma_x$ gives $F_\perp \sin\theta = mr\omega^2 = ml\sin\theta\omega^2$ and

$$F_\perp = ml\omega^2 = \left(\frac{1900\text{ N}}{9.80\text{ m/s}^2}\right)(15.0\text{ m})(0.838\text{ rad/s})^2 = 2.04 \times 10^3\text{ N}.$$

$$\Delta l = \left(\frac{2.04 \times 10^3\text{ N}}{1900\text{ N}}\right)(0.18\text{ mm}) = 0.19\text{ mm}$$

Reflect: $\sum F_y = ma_y$ gives $F_\perp \cos\theta = mg$ and $\cos\theta = mg/F_\perp$. As ω increases $F_\perp$ increases and $\cos\theta$ becomes small. Smaller $\cos\theta$ means θ increases, so the rods move toward the horizontal as ω increases.

11.71. Set Up: $\Delta V = -V_0 \dfrac{\Delta p}{B}$. Density is $\rho = \dfrac{m}{V}$. For water, $B = 2.2 \times 10^9$ Pa and air pressure at the surface of the earth is $p_0 = 1.01 \times 10^5$ Pa. $\Delta p = 92p_0 - p_0 = 91p_0$.

Solve: (a) On earth $\rho_E = m/V_0$. On Venus this same mass m of water has volume

$$V = V_0 + \Delta V = V_0\left(1 - \frac{\Delta p}{B}\right) = V_0\left(1 - \frac{91(1.01 \times 10^5\text{ Pa})}{2.2 \times 10^9\text{ Pa}}\right) = 0.9958V_0.$$

On Venus,

$$\rho_V = \frac{m}{V} = \frac{m}{0.9958V_0} = 1.004\frac{m}{V_0} = 1.004\rho_E.$$

$\rho_E = 1000\text{ kg/m}^3$ so $\rho_V = 1004\text{ kg/m}^3$ and $\rho_V/\rho_E = 100.4\%$. The density increases by 0.4%.
(b) The difference is very small and not noticeable.
Reflect: The increase in pressure decreases V and increases density.

MECHANICAL WAVES AND SOUND

Problems 1, 7, 9, 13, 15, 19, 21, 23, 29, 31, 35, 39, 45, 47, 49, 53, 55, 57, 59, 61, 65, 67, 75, 77, 79, 81, 83

Solutions to Problems

12.1. Set Up: $v = f\lambda$

Solve: **(a)** $v = 344$ m/s. For $f = 20{,}000$ Hz, $\lambda = \dfrac{v}{f} = \dfrac{344 \text{ m/s}}{20{,}000 \text{ Hz}} = 1.7$ cm. For $f = 20$ Hz,

$$\lambda = \frac{v}{f} = \frac{344 \text{ m/s}}{20 \text{ Hz}} = 17 \text{ m}.$$

The range of wavelengths is 1.7 cm to 17 m.
(b) $v = c = 3.00 \times 10^8$ m/s. For $\lambda = 700$ nm,

$$f = \frac{c}{\lambda} = \frac{3.00 \times 10^8 \text{ m/s}}{700 \times 10^{-9} \text{ m}} = 4.3 \times 10^{14} \text{ Hz}.$$

For $\lambda = 400$ nm,

$$f = \frac{c}{\lambda} = \frac{3.00 \times 10^8 \text{ m/s}}{400 \times 10^{-9} \text{ m}} = 7.5 \times 10^{14} \text{ Hz}.$$

The range of frequencies for visible light is 4.3×10^{14} Hz to 7.5×10^{14} Hz.
(c) $v = 344$ m/s. $\lambda = \dfrac{v}{f} = \dfrac{344 \text{ m/s}}{23 \times 10^3 \text{ Hz}} = 1.5$ cm
(d) $v = 1480$ m/s. $\lambda = \dfrac{v}{f} = \dfrac{1480 \text{ m/s}}{23 \times 10^3 \text{ Hz}} = 6.4$ cm
Reflect: For a given v, larger f corresponds to smaller λ. For the same f, λ increases when v increases.

12.7. Set Up: $\mu = 0.0550$ kg/m. $F_\perp = mg$, the weight of the hanging mass. The frequency of the waves is 120 Hz. $v = f\lambda$.
Solve: **(a)** $v = \sqrt{\dfrac{F_\perp}{\mu}} = \sqrt{\dfrac{(1.50 \text{ kg})(9.80 \text{ m/s}^2)}{0.0550 \text{ kg/m}}} = 16.3$ m/s
(b) $\lambda = \dfrac{v}{f} = \dfrac{16.3 \text{ m/s}}{120 \text{ Hz}} = 0.136$ m
(c) Doubling the mass increases both the wave speed and the wavelength by a factor of $\sqrt{2}$; $v = 23.1$ m/s and $\lambda = 0.192$ m.
Reflect: The speed of the waves on the rope is much less than the speed of sound in air.

12.9. Set Up: $v = \sqrt{\dfrac{F_T}{\mu}} = \sqrt{\dfrac{F_T L}{m}}$, where m is the mass of the wire.
Solve: $F_T = Mg$. $V = \sqrt{\dfrac{MgL}{m}} = \sqrt{\dfrac{Mg}{\mu}}$

(a) $M \to 2M$ gives $V_{new} = \sqrt{2}\,V$

(b) μ is the same since an identical wire twice as long will have twice the mass. $V_{new} = V$

(c) $m \to 3m$; $V_{new} = V/\sqrt{3}$

(d) $L \to 2L$ while m stays the same; $V_{new} = \sqrt{2}\,V$

(e) $L \to 2L$ while m stays the same and $M \to 10M$; $V_{new} = \sqrt{2(10)}\,V = 2\sqrt{5}\,V$

Reflect: The speed increases when the tension in the wire increases and the speed decreases when the mass per unit length μ increases.

12.13. Set Up: For a wave traveling in the $+x$ direction,

$$y(x, t) = A\sin 2\pi\left(\frac{t}{T} - \frac{x}{\lambda}\right).$$

$v = f\lambda$ and $v = \sqrt{\dfrac{F_T}{\mu}}$.

Solve: By comparing to the general equation, $\dfrac{2\pi}{T} = 415 \text{ s}^{-1}$ and $\dfrac{2\pi}{\lambda} = 44.9 \text{ m}^{-1}$. $T = 0.0151$ s and $\lambda = 0.140$ m.

$$v = f\lambda = \frac{\lambda}{T} = \frac{0.140 \text{ m}}{0.0151 \text{ s}} = 9.27 \text{ m/s}.$$

$$\mu = \frac{F_T}{v^2} = \frac{4.00 \text{ N}}{(9.27 \text{ m/s})^2} = 4.65 \times 10^{-2} \text{ kg/m} = 0.465 \text{ g/cm}$$

Reflect: A 1-meter length of the string has a mass of 0.0465 kg and a weight of about 1.6 ounces. This is a reasonable value.

12.15. Set Up: The waves obey the principle of superposition: the resultant displacement of the string is the algebraic sum of the displacements due to each wave. Each wave travels 1 cm each second.

Solve: The locations of each wave pulse and the resultant displacement of the string is shown at each time in Figures 12.15 a–d.

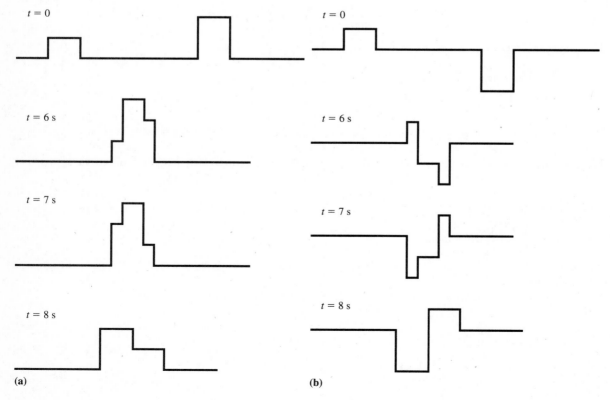

(a)   (b)

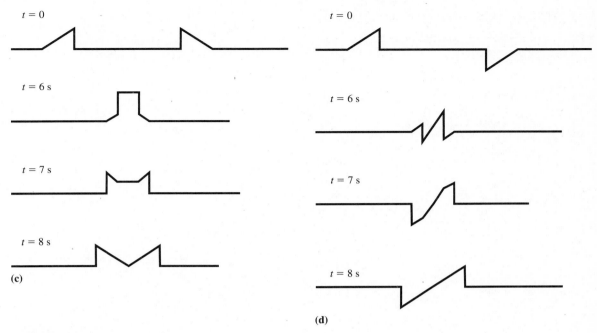

Figure 12.15

Reflect: When the pulses displace the string in the same direction they reinforce each other and give a larger pulse and when the pulses displace the string in opposite directions they tend to cancel and give a smaller pulse.

12.19. Set Up: For the fundamental, $f_1 = \dfrac{v}{2x}$, where x is the length of the portion of the string that is free to vibrate. The wave speed v depends on the tension and linear mass density of the string, so it is the same no matter where the finger is placed.

(a) For $x = L = 0.600$ m, $f_1 = 440$ Hz. $v = 2xf_1 = 2(0.600 \text{ m})(440 \text{ Hz}) = 528 \text{ m/s}$. Then for $f_1 = 587$ Hz,

$$x = \frac{v}{2f_1} = \frac{528 \text{ m/s}}{2(587 \text{ Hz})} = 0.450 \text{ m} = 45.0 \text{ cm}$$

(b) For $f_1 = 392$ Hz,

$$\lambda = \frac{v}{2f_1} = 0.673 \text{ m}.$$

The maximum length of the vibrating string, the distance between the bridge and the upper end of the fingerboard, is 0.600 m, so this is not possible. A G_4 note cannot be played without retuning.

Reflect: The speed of the waves on the string is larger than the speed of sound in air.

12.21. Set Up: $T = 1/f$ is the time for one cycle. For $f = 370$ Hz, $T = 2.70 \times 10^{-3}$ s and this is the distance along the time axis occupied by one cycle of this wave. For $f = 740$ Hz, $T = 1.35 \times 10^{-3}$ s and for $f = 1110$ Hz, $T = 0.901 \times 10^{-3}$ s. The net displacement is the algebraic sum of the displacements due to each wave.

Solve: The three waves and their sum are shown in Figure 12.21.

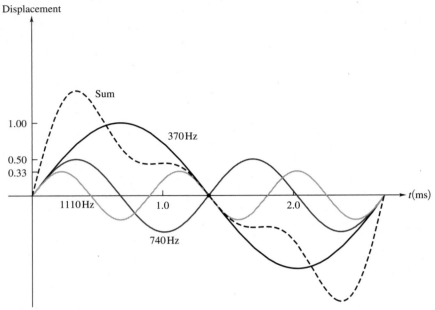

Figure 12.21

Reflect: The net sound wave is periodic, with period 2.70 ms, but it is not sinusoidal.

12.23. Set Up: An open end is a displacement antinode. A closed end is a displacement node. Adjacent nodes are a distance $\lambda/2$ apart. Adjacent antinodes are also a distance $\lambda/2$ apart, and the node to antinode distance is $\lambda/4$. Let x be the distance from the left-hand end of the pipe.

Solve: (a) There is an antinode at $x = 0$. The first node is at $x = \lambda/4$ and each successive node is $\lambda/2$ farther to the right. Each successive overtone adds one node.

fundamental: $\lambda_1 = 2L = 2.40$ m. $\lambda_1/4 = 0.60$ m. There is a node at $x = 0.60$ m.

1ˢᵗ overtone: $\lambda_2 = L = 2.40$ m. $\lambda_2/4 = 0.30$ m. There are a nodes at $x = 0.30$ m, 0.90 m.

2ⁿᵈ overtone: $\lambda_3 = 2L/3 = 0.80$ m. $\lambda_3/4 = 0.20$ m. There are a nodes at $x = 0.20$ m, 0.60 m, and 1.00 m.

(b) There now is a node at $x = 0$. Successive nodes are $\lambda/2$ farther to the right.

fundamental: $\lambda_1 = 4L = 4.80$ m. $\lambda_1/2 = 2.40$ m. There only a node is at $x = 0$.

1ˢᵗ overtone: $\lambda_3 = 4L/3 = 1.60$ m. $\lambda_3/2 = 0.80$ m. There are a nodes at $x = 0$ and $x = 0.80$ m.

2ⁿᵈ overtone: $\lambda_5 = 4L/5 = 0.96$ m. $\lambda_5/2 = 0.48$ m. There are a nodes at $x = 0$, 0.48 m, and 0.96 m.

Reflect: In each case the fundamental has one node, the 1ˢᵗ overtone has two nodes, and the 2ⁿᵈ overtone has three nodes.

12.29. Set Up: For a stopped pipe, $f_1 = \dfrac{v}{4L}$. $f_1 = 220$ Hz

Solve: $L = \dfrac{v}{4f_1} = \dfrac{344 \text{ m/s}}{4(220 \text{ Hz})} = 39.1$ cm. This result is a reasonable value for the mouth to diaphragm distance for a typical adult.

Reflect: 1244 Hz is not an integer multiple of the fundamental frequency of 220 Hz, it is 5.65 times the fundamental. The production of sung notes is more complicated than harmonics of an air column of fixed length.

12.31. Set Up: For an open pipe the first overtone has frequency $f_2 = 2f_1 = 2\dfrac{v}{2L} = \dfrac{v}{L}$. For a stopped pipe the first overtone has frequency $f_3 = 3f_1 = \dfrac{3v}{4L}$. The mouthpiece acts as a closed end.

Solve: **(a)** $f_3 = \dfrac{3v}{4L} = \dfrac{3(344 \text{ m/s})}{4(3.7 \text{ m})} = 70 \text{ Hz}$

(b) A pipe closed at both ends has the same overtone frequencies as a pipe open at both ends (Problem 12.28).

$$f_2 = \frac{v}{L} = \frac{344 \text{ m/s}}{3.7 \text{ m}} = 93 \text{ Hz}$$

Reflect: The fundamental frequency for a stopped pipe is lower by a factor of two than the fundamental frequency for a pipe of the same length but closed at both ends. But for the stopped pipe the first overtone is the third harmonic and the first overtone is lower than that of the other pipe only by a factor of $4/3$.

12.35. Set Up: One speaker is 4.50 m from the microphone and the other is 4.03 m from the microphone, so the path difference is 0.42 m. For constructive interference, the path difference is an integer number of wavelengths. For destructive interference, the path difference is a half-integer number of wavelengths. $f = v/\lambda$

Solve: **(a)** $\lambda = 0.42$ m gives $f = \dfrac{v}{\lambda} = 820 \text{ Hz}$;

$2\lambda = 0.42$ m gives $\lambda = 0.21$ m and $f = \dfrac{v}{\lambda} = 1640 \text{ Hz}$;

$3\lambda = 0.42$ m gives $\lambda = 0.14$ m and $f = \dfrac{v}{\lambda} = 2460 \text{ Hz}$, and so on.

The frequencies for constructive interference are $n(820 \text{ Hz})$, $n = 1, 2, 3, \ldots$.

(b) $\lambda/2 = 0.42$ m gives $\lambda = 0.84$ m and $f = \dfrac{v}{\lambda} = 410 \text{ Hz}$;

$3\lambda/2 = 0.42$ m gives $\lambda = 0.28$ m and $f = \dfrac{v}{\lambda} = 1230 \text{ Hz}$;

$5\lambda/2 = 0.42$ m gives $\lambda = 0.168$ m and $f = \dfrac{v}{\lambda} = 2050 \text{ Hz}$, and so on.

The frequencies for destructive interference are $(2n + 1)(410 \text{ Hz})$, $n = 0, 1, 2, \ldots$.
Reflect: The frequencies for constructive interference lie midway between the frequencies for destructive interference.

12.39. Set Up: $v = f\lambda$
Solve: **(a)** The boundary surface between the two materials cannot create or destroy waves; the number entering one material per unit time must equal the number leaving the other material in the same time. Since $v = f\lambda$ and f doesn't change, if v changes then the wavelength λ must change.

(b) $f = \dfrac{v}{\lambda}$ is the same, so $\dfrac{v_{\text{air}}}{\lambda_{\text{air}}} = \dfrac{v_{\text{water}}}{\lambda_{\text{water}}}$.

$$\lambda_{\text{water}} = \lambda_{\text{air}}\left(\frac{v_{\text{water}}}{v_{\text{air}}}\right) = (2.50 \text{ m})\left(\frac{1498 \text{ m/s}}{344 \text{ m/s}}\right) = 10.9 \text{ m}$$

Reflect: The wavelength is larger in the material where the wave speed is larger.

12.45. Set Up: Example 12.10 shows that $\beta_2 - \beta_1 = (10 \text{ dB}) \log\left(\dfrac{I_2}{I_1}\right)$. $\dfrac{I_2}{I_1} = \dfrac{r_1^2}{r_2^2}$

Solve: $\Delta\beta = +40.0 \text{ dB}$. $\log\left(\dfrac{I_2}{I_1}\right) = 4.00$ and $\dfrac{I_2}{I_1} = 1.00 \times 10^4$

$$r_2 = r_1\sqrt{\frac{I_1}{I_2}} = (15.0 \text{ m})\sqrt{\frac{1}{1.00 \times 10^4}} = 15.0 \text{ cm}$$

Reflect: A change of 10^2 in distance gives a change of 10^4 in intensity. Our analysis assumes that the sound spreads from the source uniformly in all directions.

12.47. Set Up: Intensity is energy per unit time per unit area. $\beta = 10\log\left(\dfrac{I}{I_0}\right)$, with $I_0 = 1 \times 10^{-12}$ W/m². The area of the eardrum is $A = \pi r^2$, with $r = 4.2 \times 10^{-3}$ m. Part (b) of Problem 12.42 gave $v = 0.074$ mm/s.

Solve: (a) $\beta = 110$ dB gives $11.0 = \log\left(\dfrac{I}{I_0}\right)$ and $I = (10^{11})I_0 = 0.100$ W/m².

$$E = IAt = (0.100 \text{ W/m}^2)\pi(4.2 \times 10^{-3}\text{ m})^2(1\text{ s}) = 5.5\ \mu\text{J}$$

(b) $K = \frac{1}{2}mv^2$ so

$$v = \sqrt{\frac{2K}{m}} = \sqrt{\frac{2(5.5 \times 10^{-6}\text{ J})}{2.0 \times 10^{-6}\text{ kg}}} = 2.3 \text{ m/s}.$$

This is about 31,000 times faster than the speed in Problem 12.46b.

Reflect: Even though the sound wave intensity level is very high, the rate at which energy is delivered to the eardrum is very small, because the area of the eardrum is very small.

12.49. Set Up: Example 12.9 shows that $\beta_2 - \beta_1 = (10\text{ dB})\log\left(\dfrac{I_2}{I_1}\right)$.

Solve: (a) $\Delta\beta = 5.00$ dB gives $\log\left(\dfrac{I_2}{I_1}\right) = 0.5$ and $\dfrac{I_2}{I_1} = 10^{0.5} = 3.16$

(b) $\dfrac{I_2}{I_1} = 100$ gives $\Delta\beta = 10\log(100) = 20$ dB

(c) $\dfrac{I_2}{I_1} = 2$ gives $\Delta\beta = 10\log 2 = 3.0$ dB

Reflect: Every doubling of the intensity increases the decibel level by 3.0 dB.

12.53. Set Up: $f_{beat} = f_1 - f_2$. 1 rpm = 60 Hz

Solve: (a) 575 rpm = 9.58 Hz. The frequency of the other propeller differs by 2.0 Hz, so the frequency of the other propeller is either 11.6 Hz or 7.6 Hz. These frequencies correspond to 696 rpm or 456 rpm.

(b) When the speed and rpm of the second propeller is increased the beat frequency increases, so the frequency of the second propeller moves farther from the frequency of the first and the second propeller is turning at 696 rpm.

Reflect: If the frequency of the second propeller was 7.6 Hz then it would have moved closed to the frequency of the first when its frequency was increased and the beat frequency would have decreased.

12.55. Set Up: The positive direction is from listener to source.

Solve: (a) $v_S = -25.0$ m/s. $\lambda = \dfrac{v + v_S}{f_S} = \dfrac{344 \text{ m/s} - 25.0 \text{ m/s}}{400 \text{ Hz}} = 0.798$ m.

(b) $v_S = +25.0$ m/s. $\lambda = \dfrac{344 \text{ m/s} + 25.0 \text{ m/s}}{400 \text{ Hz}} = 0.922$ m

(c) $v_S = -25.0$ m/s. $f_L = \left(\dfrac{v}{v + v_S}\right)f_S = \left(\dfrac{344 \text{ m/s}}{344 \text{ m/s} - 25.0 \text{ m/s}}\right)(400 \text{ Hz}) = 431$ Hz

(d) $v_S = +25.0$ m/s. $f_L = \left(\dfrac{344 \text{ m/s}}{344 \text{ m/s} + 25.0 \text{ m/s}}\right)(400 \text{ Hz}) = 373$ Hz

Reflect: If the train were at rest, the wavelength of the sound from its whistle would be

$$\frac{344 \text{ m/s}}{400 \text{ Hz}} = 0.860 \text{ m}.$$

When the listener is in front of the source, the source is approaching. The frequency is increases and the wavelength is decreased. When the listener is behind the source, the source is moving away. The frequency is decreased and the wavelength is increased.

12.57. Set Up: The positive direction is from listener to source. $f_S = 1200$ Hz. $f_L = 1240$ Hz.

Solve: $f_L = \left(\dfrac{v + v_L}{v + v_S}\right)f_S.$ $v_L = 0.$ $v_S = -25.0$ m/s. $f_L = \left(\dfrac{v}{v + v_S}\right)f_S$ gives

$$v = \frac{v_S f_L}{f_S - f_L} = \frac{(-25 \text{ m/s})(1240 \text{ Hz})}{1200 \text{ Hz} - 1240 \text{ Hz}} = 780 \text{ m/s}.$$

Reflect: $f_L > f_S$ since the source is approaching the listener.

12.59. Set Up: The positive direction is from the listener to the source. (a) The wall is the listener. $v_S = -30$ m/s. $v_L = 0.$ $f_L = 600$ Hz. (b) The wall is the source and the car is the listener. $v_S = 0.$ $v_L = +30$ m/s. $f_S = 600$ Hz.

Solve: (a) $f_L = \left(\dfrac{v + v_L}{v + v_S}\right)f_S.$

$$f_S = \left(\frac{v + v_S}{v + v_L}\right)f_L = \left(\frac{344 \text{ m/s} - 30 \text{ m/s}}{344 \text{ m/s}}\right)(600 \text{ Hz}) = 548 \text{ Hz}$$

(b) $f_L = \left(\dfrac{v + v_L}{v + v_S}\right)f_S = \left(\dfrac{344 \text{ m/s} + 30 \text{ m/s}}{344 \text{ m/s}}\right)(600 \text{ Hz}) = 652 \text{ Hz}$

Reflect: Since the singer and wall are moving toward each other the frequency received by the wall is greater than the frequency sung by the soprano, and the frequency she hears from the reflected sound is larger still.

12.61. Set Up:

$$f_L = \left(\frac{v + v_L}{v + v_S}\right)f_S.$$

To treat the reflection of the waves use Problem-Solving Strategy 3 in Section 12.12.

Solve: (a) Since the frequency is increased the moving car must be approaching the police car. Let v_c be the speed of the moving car. The speed v_p of the police car is zero. First consider the moving car as the listener, as shown in Figure 12.61a:

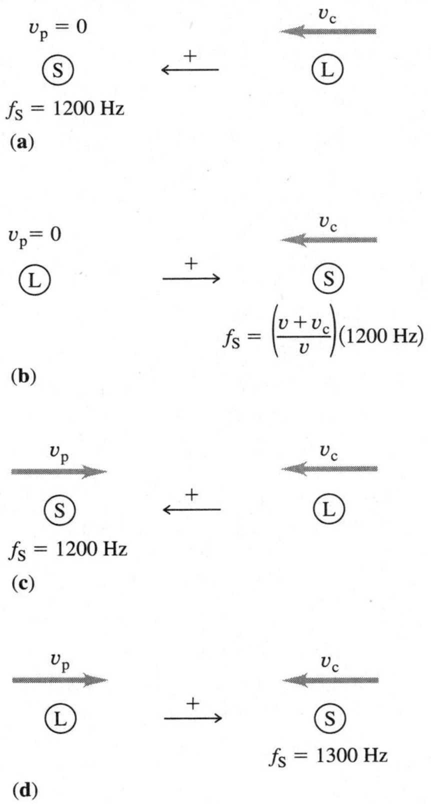

$v_p = 0$

S

$f_S = 1200$ Hz

v_c

L

$+$

(a)

$v_p = 0$

L

v_c

S

$+$

$f_S = \left(\dfrac{v + v_c}{v}\right)(1200 \text{ Hz})$

(b)

v_p

S

$f_S = 1200$ Hz

v_c

L

$+$

(c)

v_p

L

v_c

S

$+$

$f_S = 1300$ Hz

(d)

Figure 12.61

$$f_L = \left(\frac{v + v_L}{v + v_S}\right)f_S = \left(\frac{v + v_c}{v}\right)(1200 \text{ Hz})$$

Then consider the moving car as the source and the police car as the listener (Figure 12.61b):

$$f_L = \left(\frac{v + v_L}{v + v_S}\right)f_S \text{ gives } 1250 \text{ Hz} = \left(\frac{v}{v - v_c}\right)\left(\frac{v + v_c}{v}\right)(1200 \text{ Hz}).$$

Solving for v_c gives

$$v_c = \left(\frac{50}{2450}\right)v = \left(\frac{50}{2450}\right)(344 \text{ m/s}) = 7.02 \text{ m/s}$$

(b) Repeat the calculation of part (a), but now $v_p = 20.0$ m/s, toward the other car.
Waves received by the car (Figure 12.61c):

$$f_L = \left(\frac{v + v_c}{v - v_p}\right)f_S = \left(\frac{344 \text{ m/s} + 7 \text{ m/s}}{344 \text{ m/s} - 20 \text{ m/s}}\right)(1200 \text{ Hz}) = 1300 \text{ Hz}$$

Waves reflected by the car and received by the police car (Figure 12.61d):

$$f_L = \left(\frac{v + v_p}{v - v_c}\right)f_S = \left(\frac{344 \text{ m/s} + 20 \text{ m/s}}{344 \text{ m/s} - 7 \text{ m/s}}\right)(1300 \text{ Hz}) = 1404 \text{ Hz}$$

Reflect: The cars move toward each other with a greater relative speed in (b) and the increase in frequency is much larger there.

12.65. Set Up: The formula for the Doppler shift for light is $f_L = \sqrt{\dfrac{c - v}{c + v}}f_S$. The sign convention is that v is negative when the source and listener are approaching, so $v = -2.0 \times 10^7$ m/s. $c = f\lambda$
Solve: The astronomer on earth measures f_L.

$$f_L = \frac{c}{\lambda_L} = \frac{3.00 \times 10^8 \text{ m/s}}{523.4 \times 10^{-9} \text{ m}} = 5.732 \times 10^{14} \text{ Hz}$$

The observer at rest with respect to the star measures f_S and λ_S.

$$f_S = \sqrt{\frac{c + v}{c - v}}f_L = \sqrt{\frac{3.00 \times 10^8 \text{ m/s} - 2.0 \times 10^7 \text{ m/s}}{3.00 \times 10^8 \text{ m/s} + 2.0 \times 10^7 \text{ m/s}}}(5.732 \times 10^{14} \text{ Hz}) = 5.362 \times 10^{14} \text{ Hz}$$

$$\lambda_S = \frac{c}{f_S} = \frac{3.00 \times 10^8 \text{ m/s}}{5.362 \times 10^{14} \text{ Hz}} = 559.5 \text{ nm}$$

Reflect: The source and observer are approaching so $f_L > f_S$ and $\lambda_L < \lambda_S$. Note that v is the relative velocity of the source and receiver.

12.67. Set Up: The heart wall first acts as the listener and then as the source. The positive direction is from listener to source. The heart wall is moving toward the receiver so the Doppler effect increases the frequency and the final frequency received, f_{L2}, is greater than the source frequency, f_{S1}. $f_{L2} - f_{S1} = 85$ Hz.
Solve: Heart wall receives the sound: $f_S = f_{S1}$. $f_L = f_{L1}$. $v_S = 0$. $v_L = -v_{wall}$.

$$f_L = \left(\frac{v + v_L}{v + v_S}\right)f_S \text{ gives } f_{L1} = \left(\frac{v - v_{wall}}{v}\right)f_{S1}.$$

Heart wall emits the sound: $f_{S2} = f_{L1}$. $v_S = +v_{wall}$. $v_L = 0$.

$$f_{L2} = \left(\frac{v}{v + v_{wall}}\right)f_{S2} = \left(\frac{v}{v + v_{wall}}\right)\left(\frac{v - v_{wall}}{v}\right)f_{S1} = \left(\frac{v - v_{wall}}{v + v_{wall}}\right)f_{S1}.$$

$$f_{L2} - f_{S1} = \left(1 - \frac{v - v_{wall}}{v + v_{wall}}\right)f_{S1} = \left(\frac{2v_{wall}}{v + v_{wall}}\right)f_{S1}.$$

$$v_{wall} = \frac{(f_{L2} - f_{S1})v}{2f_{S1} - (f_{L2} - f_{S1})}.$$

$$f_{S1} \gg f_{L2} - f_{S1} \text{ and}$$

$$v_{wall} = \frac{(f_{L2} - f_{S1})v}{2f_{S1}} = \frac{(85 \text{ Hz})(1500 \text{ m/s})}{2(2.00 \times 10^6 \text{ Hz})} = 0.0319 \text{ m/s} = 3.19 \text{ cm/s}.$$

Reflect: $f_{S1} = 2.00 \times 10^6$ Hz and $f_{L2} - f_{S1} = 85$ Hz, so the approximation we made is very accurate. Within this approximation, the frequency difference between the reflected and transmitted waves is directly proportional to the speed of the heart wall.

12.75. Set Up: The whale acts as the listener and then as the source. The direction from the listener to the source is positive. $f_S = 22.0 \times 10^3$ Hz.

Solve: (a) $\lambda = \frac{v}{f} = \frac{1482 \text{ m/s}}{22.0 \times 10^3 \text{ Hz}} = 0.0674$ m

(b) whale receives the sound: $v_S = 0$, $v_L = +v_{whale}$. $f_{L1} = f_L$.

$$f_{L1} = \left(\frac{v + v_L}{v + v_S}\right)f_S = \left(\frac{v + v_{whale}}{v}\right)f_S$$

whale transmits the sound: $v_S = -v_{whale}$, $v_L = 0$. $f_S = f_{S2} = f_{L1}$. $f_{L2} = f_L$.

$$f_{L2} = \left(\frac{v}{v - v_{whale}}\right)\left(\frac{v + v_{whale}}{v}\right)f_S = \left(\frac{v + v_{whale}}{v - v_{whale}}\right)f_S.$$

$$\Delta f = f_{L2} - f_S = \left(\frac{v + v_{whale}}{v - v_{whale}} - 1\right)f_S = \left(\frac{2v_{whale}}{v - v_{whale}}\right)f_S = \left(\frac{2[4.95 \text{ m/s}]}{1482 \text{ m/s} - 4.95 \text{ m/s}}\right)(22.0 \times 10^3 \text{ Hz}) = 147 \text{ Hz}$$

Reflect: The speed of the whale is much less than the speed of the waves and the fractional shift in frequency is very small.

12.77. Set Up: The distance between crests is λ. In front of the source $\lambda = \frac{v - v_S}{f_S}$ and behind the source $\lambda = \frac{v + v_S}{f_S}$.

Solve: (a) $T = 1.6$ s and $f_S = 1/T = 0.625$ Hz. $\lambda = \frac{v - v_S}{f_S}$ gives

$$v_S = v - \lambda f_S = 0.32 \text{ m/s} - (0.12 \text{ m})(0.625 \text{ Hz}) = 0.25 \text{ m/s}.$$

(b) $\lambda = \frac{v + v_S}{f_S} = \frac{0.32 \text{ m/s} + 0.25 \text{ m/s}}{0.625 \text{ Hz}} = 0.91$ m

Reflect: If the duck was held at rest but still paddled its feet, it would produce waves of wavelength

$$\lambda = \frac{0.32 \text{ m/s}}{0.625 \text{ Hz}} = 0.51 \text{ m}.$$

In front of the duck the wavelength is decreased and behind the duck the wavelength is increased. The speed of the duck is 78% of the wave speed, so the Doppler effects are large.

12.79. Set Up: Apply $f_L = \left(\dfrac{v + v_L}{v + v_S}\right) f_S$ to the sound from each wire to find expressions for f_{LA} and f_{LB}, the frequencies of the sounds from wires A and B as detected by the bee. $f_{LA} - f_{LB} = 15$ Hz. $v = 344$ m/s.

Solve: *wire A* (Figure 12.79a) $f_{LA} = \left(\dfrac{v + 8.0 \text{ m/s}}{v}\right) f_S = 1.0233 f_S$

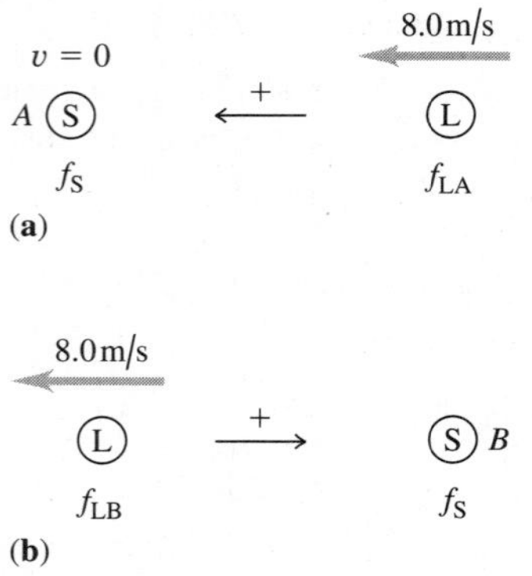

(a)

(b)

Figure 12.79

wire B (Figure 12.79b) $f_{LB} = \left(\dfrac{v - 8.0 \text{ m/s}}{v}\right) f_S = 0.9767 f_S$

$f_{LA} - f_{LB} = 15$ Hz so $1.0233 f_S - 0.9767 f_S = 15$ Hz and $f_S = 320$ Hz

Reflect: $f_{LA} > f_S$ because the bee is flying toward A and $f_{LB} < f_S$ because the bee is flying away from B.

12.81. Set Up: The greatest frequency shift from the Doppler effect occurs when one speaker is moving away and one is moving toward the person. The speakers have speed $v_0 = r\omega$, where $r = 0.75$ m. $f_L = \left(\dfrac{v + v_L}{v + v_S}\right) f_S$, with the positive direction from the listener to the source. $v = 344$ m/s

Solve: (a) $f = \dfrac{v}{\lambda} = \dfrac{344 \text{ m/s}}{0.313 \text{ m}} = 1100$ Hz.

$$\omega = (75 \text{ rpm})\left(\frac{2\pi \text{ rad}}{1 \text{ rev}}\right)\left(\frac{1 \text{ min}}{60 \text{ s}}\right) = 7.85 \text{ rad/s}$$

and $v_0 = (0.75 \text{ m})(7.85 \text{ rad/s}) = 5.89$ m

Speaker A, moving toward the listener:

$$f_{LA} = \left(\frac{v}{v - 5.89 \text{ m/s}}\right)(1100 \text{ Hz}) = 1119 \text{ Hz}.$$

Speaker B, moving toward the listener:

$$f_{LB} = \left(\frac{v}{v + 5.89 \text{ m/s}}\right)(1100 \text{ Hz}) = 1081 \text{ Hz}$$

$$f_{\text{beat}} = f_1 - f_2 = 1119 \text{ Hz} - 1081 \text{ Hz} = 38 \text{ Hz}$$

(b) A person can her individual beats only up to about 7 Hz and this beat frequency is much larger than that.

Reflect: As the turntable rotates faster the beat frequency at this position of the speakers increases.

12.83. Set Up: $f_1 = \dfrac{v}{2L}$ and $v = \sqrt{\dfrac{TL}{m}}$ so $f_1 = \dfrac{1}{2}\sqrt{\dfrac{T}{Lm}}$, where m is the mass of the wire.

Solve: With the 100 kg weight,

$$f_1 = \frac{1}{2}\sqrt{\frac{(100 \text{ kg})(9.80 \text{ m/s}^2)}{(1.40 \text{ m})(4.50 \times 10^{-3} \text{ kg})}} = 197 \text{ Hz}.$$

Adding mass increases the frequency, so add mass until $f_1' = 202$ Hz.

$$T = Lm(2f_1')^2 = (1.40 \text{ m})(4.50 \times 10^{-3} \text{ kg})(4)(202 \text{ Hz})^2 = 1028 \text{ N}$$

The total mass suspended from the wire is 105 kg and 5 kg has been added.

Reflect: The frequency is proportional to the square root of the tension, so a 5% increase in tension produces a 2.5% increase in frequency. $v = 344$ m/s.

13

FLUID MECHANICS

Problems 5, 7, 9, 13, 19, 21, 23, 25, 27, 29, 31, 33, 35, 41, 45, 49, 51, 55, 59, 61, 63, 67, 69, 71

Solutions to Problems

13.5. Set Up: $\rho = m/V$. For a sphere, $V = \frac{4}{3}\pi r^3$.

Solve: (a) The mass of the nugget is $\dfrac{\$1.00 \times 10^6}{\$426.6/\text{ounce}} = (2344 \text{ ounces})\left(\dfrac{31.1035 \text{ g}}{1 \text{ ounce}}\right) = 7.29 \times 10^4 \text{ g}$

$$V = \frac{m}{\rho} = \frac{7.29 \times 10^4 \text{ g}}{19.3 \text{ g/cm}^3} = 3.78 \times 10^3 \text{ cm}^3$$

$$r = \left(\frac{3V}{4\pi}\right)^{1/3} = \left[\frac{3(3.78 \times 10^3 \text{ cm}^3)}{4\pi}\right]^{1/3} = 9.66 \text{ cm}$$

and its diameter is $2r = 19.3 \text{ cm}$

(b) The platinum nugget would have mass

$$m = \rho V = (21.4 \text{ g/cm}^3)(3.78 \times 10^3 \text{ cm}^3) = 8.09 \times 10^4 \text{ g} = 2.60 \times 10^3 \text{ troy ounces}$$

and would be worth 2.29 million dollars.

Reflect: The platinum nugget is worth more because a nugget of the same size has more mass, since platinum's density is greater, and because it is worth more per ounce. The value if platinum is larger than the value if gold by $(21.4/19.3)(879.00/426.60)$.

13.7. Set Up: $\rho = m/V = m/L^3$

Solve: (a) $m = \rho L^3$ so $\rho_1 L_1^3 = \rho_2 L_2^3$ and $\rho_2 = \rho_1\left(\dfrac{L_1}{L_2}\right)^3 = \rho\left(\dfrac{L_1}{L_1/2}\right)^3 = 8\rho$

(b) $L_2 = L_1\left(\dfrac{\rho_1}{\rho_2}\right)^{1/3} = L\left(\dfrac{\rho_1}{3\rho_1}\right)^{1/3} = L/3^{1/3}$

Reflect: If the volume is decreased while the mass is kept the same, then the density increases.

13.9. Set Up: $\rho = m/V$. $m_{\text{mix}} = m_1 + m_2$ and $V_{\text{mix}} = V_1 + V_2$

Solve: (a) $m_1 = m_2 = m$ so $m_{\text{mix}} = 2m$. $V_1 = \dfrac{m}{\rho_1}$, $V_2 = \dfrac{m}{\rho_2}$ so $V_{\text{mix}} = m\left(\dfrac{1}{\rho_1} + \dfrac{1}{\rho_2}\right) = m\left(\dfrac{\rho_1 + \rho_2}{\rho_1\rho_2}\right)$ and

$$\rho_{\text{mix}} = \frac{m_{\text{mix}}}{V_{\text{mix}}} = \frac{2m}{m\left(\dfrac{\rho_1 + \rho_2}{\rho_1\rho_2}\right)} = \frac{2\rho_1\rho_2}{\rho_1 + \rho_2}\rho_{\text{av}}$$

$\rho_{\text{av}} = \dfrac{\rho_1 + \rho_2}{2}$.

In general ρ_{mix} is not equal to ρ_{av} and this proves that the answer is no. For example, if $\rho_1 = 1.00 \text{ g/cm}^3$ and $\rho_2 = 2.00 \text{ g/cm}^3$, then $\rho_{\text{mix}} = 1.33 \text{ g/cm}^3$ and $\rho_{\text{av}} = 1.50 \text{ g/cm}^3$, which is different,

(b) $V_1 = V_2 = V$ so $V_{mix} = 2V$. $m_1 = \rho_1 V$, $m_2 = \rho_2 V$ so $m_{mix} = (\rho_1 + \rho_2)V$

$$\rho_{mix} = \frac{m_{mix}}{V_{mix}} = \frac{(\rho_1 + \rho_2)V}{2V} = \frac{\rho_1 + \rho_2}{2},$$

which is the same as the average of the two densities. This proves that the answer is yes.

Reflect: If $\rho_1 = \rho_2 = \rho$ then in (a) both ρ_{mix} and ρ_{av} equal ρ.

13.13. Set Up: The density of seawater is $1.03 \times 10^3 \text{ kg/m}^3$.

Solve: (a) $p_{gauge} = p_0 - p_{atm} = \rho g h = (1.03 \times 10^3 \text{ kg/m}^3)(9.80 \text{ m/s}^2)(250 \text{ m}) = 2.52 \times 10^6 \text{ Pa}$.

(b) $F_{net} = p_{gauge} A = (2.52 \times 10^6 \text{ Pa})\pi(0.150 \text{ m})^2 = 1.78 \times 10^5 \text{ N}$

Reflect: At the surface the gauge pressure is zero. At a depth of 250 m the force due to the water pressure is very large.

13.19. Set Up: 1 torr = 1 mm Hg = 133.3 Pa. 1 atm = 1.013×10^5 Pa = 760 torr.

Solve: (a) $p = \rho g h$ so $h = \dfrac{p}{\rho g} = \dfrac{133.3 \text{ Pa}}{(1.00 \times 10^3 \text{ kg/m}^3)(9.80 \text{ m/s}^2)} = 1.36 \text{ cm}$

(b) 1 atm = 760 torr so $h = (760)(1.36 \text{ cm}) = 10.3 \text{ m}$

The height of a column of mercury at $p = 1$ atm would be only 0.76 m.

Reflect: If water is used, the height of the barometer for ordinary atmospheric pressures is impractical. Mercury is used instead because its density is larger.

13.21. Set Up: 1 atm = 1.013×10^5 Pa. The pressure difference is $\Delta p = \rho g h$.

Solve: (a) $\Delta p = 0.125$ atm = 1.27×10^4 Pa.

$$\rho = \frac{\Delta p}{gh} = \frac{1.27 \times 10^4 \text{ Pa}}{(9.80 \text{ m/s}^2)(1.50 \text{ m})} = 864 \text{ kg/m}^3$$

(b) $\Delta p = \rho g_{Mars} h = (864 \text{ kg/m}^3)(0.379)(9.80 \text{ m/s}^2)(1.50 \text{ m}) = 4.81 \times 10^3 \text{ Pa} = 0.0475 \text{ atm}$

Reflect: The pressure difference on Mars is less than it is on earth by a factor of 0.379. Our calculated density of oil is a reasonable value, somewhat less than the density of water.

13.23. Set Up: $p_0 = p_{surface} + \rho g h$ where $p_{surface}$ is the pressure at the surface of a liquid and p_0 is the pressure at a depth h below the surface.

Solve: (a) For the oil layer, $p_{surface} = p_{atm}$ and p_0 is the pressure at the oil-water interface.

$$p_0 - p_{atm} = p_{gauge} = \rho g h = (600 \text{ kg/m}^3)(9.80 \text{ m/s}^2)(0.120 \text{ m}) = 706 \text{ Pa}$$

(b) For the water layer, $p_{surface} = 706 \text{ Pa} + p_{atm}$.

$$p_0 - p_{atm} = p_{gauge} = 706 \text{ Pa} + \rho g h = 706 \text{ Pa} + (1.00 \times 10^3 \text{ kg/m}^3)(9.80 \text{ m/s}^2)(0.250 \text{ m}) = 3.16 \times 10^3 \text{ Pa}$$

Reflect: The gauge pressure at the bottom of the barrel is due to the combined effects of the oil layer and water layer. The pressure at the bottom of the oil layer is the pressure at the top of the water layer

13.25. Set Up: 1 torr = 1 mm Hg = 133.3 Pa. The pressure change with depth is $\Delta p = \rho g h$. The arm is above the heart, so the pressure in the arm is less than the pressure at the heart. On the moon, $g = 1.67 \text{ m/s}^2$.

Solve: The pressure in the arm is less by an amount

$$\rho g h = (1050 \text{ kg/m}^3)(1.67 \text{ m/s}^2)(0.25 \text{ m}) = 438 \text{ Pa} = 3.3 \text{ torr}$$

The blood pressure reading should be $\dfrac{120 - 3}{80 - 3} = \dfrac{117}{77}$.

Reflect: The changes in blood pressure for points in the body above or below the heart is much less on the moon than on Earth, since g is much less on the moon.

13.27. Set Up: Pascal's law says the pressure is the same everywhere in the hydraulic fluid, so $F_1/A_1 = F_2/A_2$. $F_1 = 100$ N and $F = (3500 \text{ kg})(9.80 \text{ m/s}^2) = 3.43 \times 10^4$ N, the weight of the car and platform that is being lifted. The volume of fluid displaced at each piston when the pistons move distances d_1 and d_2 is the same, so $d_1 A_1 = d_2 A_2$.

Solve: (a) $A = \pi r^2$ so $\dfrac{F_1}{r_1^2} = \dfrac{F_2}{r_2^2}$. $r_2 = r_1\sqrt{\dfrac{F_2}{F_1}} = (0.125 \text{ m})\sqrt{\dfrac{3.43 \times 10^4 \text{ N}}{100 \text{ N}}} = 2.32$ m and the diameter is 4.64 m.

(b) $d_1 A_1 = d_2 A_2$.

$$d_2 = d_1\frac{A_1}{A_2} = d_1\frac{\pi r_1^2}{\pi r_2^2} = (50 \text{ cm})\left(\frac{0.125 \text{ m}}{2.32 \text{ m}}\right)^2 = 1.45 \text{ mm}$$

Reflect: The work done by the man is $F_1 d_1 = (100 \text{ N})(0.50 \text{ m}) = 50$ J. The work done on the car is $F_2 d_2 = (3.43 \times 10^4 \text{ N})(1.43 \times 10^{-3} \text{ m}) = 50$ J. Energy conservation requires that these two quantities be equal, and they are equal.

13.29. Set Up: The density of aluminum is 2.7×10^3 kg/m^3. The density of water is 1.00×10^3 kg/m^3. $\rho = m/V$. The buoyant force is $F_B = \rho_{\text{water}} V_{\text{obj}} g$.

Solve: (a) $T = mg = 89$ N so $m = 9.08$ kg. $V = \dfrac{m}{\rho} = \dfrac{9.08 \text{ kg}}{2.7 \times 10^3 \text{ kg/m}^3} = 3.36 \times 10^{-3}$ m^3 = 3.4 L.

(b) When the ingot is totally immersed in the water while suspended, $T + F_B - mg = 0$.

$$F_B = \rho_{\text{water}} V_{\text{obj}} g = (1.00 \times 10^3 \text{ kg/m}^3)(3.36 \times 10^{-3} \text{ m}^3)(9.80 \text{ m/s}^2) = 32.9 \text{ N}.$$

$$T = mg - F_B = 89 \text{ N} - 32.9 \text{ N} = 56 \text{ N}.$$

Reflect: The buoyant force is equal to the difference between the apparent weight when the object is submerged in the fluid and the actual gravity force on the object.

13.31. Set Up: $F_B = \rho_{\text{fluid}} g V_{\text{sub}}$, where V_{sub} is the volume of the object that is below the fluid's surface.
Solve: (a) Floats, so the buoyant force equals the weight of the object: $F_B = mg$. Using Archimedes's principle gives $\rho_w g V = mg$ and

$$V = \frac{m}{\rho_w} = \frac{5750 \text{ kg}}{1.00 \times 10^3 \text{ kg/m}^3} = 5.75 \text{ m}^3.$$

(b) $F_B = mg$ and $\rho_w g V_{\text{sub}} = mg$. $V_{\text{sub}} = 0.80V = 4.60$ m^3, so the mass of the floating object is

$$m = \rho_w V_{\text{sub}} = (1.00 \times 10^3 \text{ kg/m}^3)(4.60 \text{ m}^3) = 4600 \text{ kg}.$$

He must throw out 5750 kg − 4600 kg = 1150 kg.
Reflect: He must throw out 20% of the boat's mass.

13.33. Set Up: $F_B = \rho_{\text{water}} V_{\text{ice}} g$. $m_{\text{ice}} = \rho_{\text{ice}} V$. $\rho_{\text{ice}} = 0.92 \times 10^3$ kg/m^3.
Solve: The ice is floating, so $F_B = (m_{\text{ice}} + m_{\text{woman}})g$. $\rho_{\text{water}} V_{\text{ice}} = \rho_{\text{ice}} V_{\text{ice}} + m_{\text{woman}}$.

$$V_{\text{ice}} = \frac{m_{\text{woman}}}{\rho_{\text{water}} - \rho_{\text{ice}}} = \frac{45.0 \text{ kg}}{1000 \text{ kg/m}^3 - 920 \text{ kg/m}^3} = 0.562 \text{ m}^3.$$

Reflect: The buoyant force must support both the slab of ice and the woman. If the slab of ice is 20 cm thick and square on its top and bottom surfaces, its top surface measures about 1.7 m on a side. The woman would also need to be concerned about the slab tipping to one side, because of unbalanced torques.

13.35. Set Up: $F_B = \rho_{\text{water}} V_{\text{obj}} g$. The net force on the sphere is zero.
Solve: (a) $F_B = (1000 \text{ kg/m}^3)(0.650 \text{ m}^3)(9.80 \text{ m/s}^2) = 6.37 \times 10^3$ N
(b) $F_B = T + mg$ and

$$m = \frac{F_B - T}{g} = \frac{6.37 \times 10^3 \text{ N} - 900 \text{ N}}{9.80 \text{ m/s}^2} = 558 \text{ kg}.$$

(c) Now $F_B = \rho_{water}V_{sub}g$, where V_{sub} is the volume of the sphere that is submerged. $F_B = mg$. $\rho_{water}V_{sub} = mg$ and

$$V_{sub} = \frac{m}{\rho_{water}} = \frac{558 \text{ kg}}{1000 \text{ kg/m}^3} = 0.558 \text{ m}^3.$$

$$\frac{V_{sub}}{V_{obj}} = \frac{0.558 \text{ m}^3}{0.650 \text{ m}^3} = 0.858 = 85.8\%$$

Reflect: When the sphere is totally submerged, the buoyant force on it is greater than its weight. When it is floating, it needs to be only partially submerged in order to produce a buoyant force equal to its weight.

13.41. Set Up: The gauge pressure $p - p_{atm}$ in a drop is related to the surface tension γ of the liquid and the radius R of the drop by $p - p_{atm} = \frac{4\gamma}{R}$. The volume of a spherical drop is $V = \frac{4}{3}\pi R^3$.

Solve: $R = \left(\frac{3V}{4\pi}\right)^{1/3}$ so $p - p_{atm} = 4\gamma\left(\frac{4\pi}{3V}\right)^{1/3}$. Same gauge pressure, so $\gamma V^{-1/3} = $ constant and $\gamma_w V_w^{-1/3} = \gamma_{ct}V_{ct}^{-1/3}$.

$$\frac{V_w}{V_{ct}} = \left(\frac{\gamma_w}{\gamma_{ct}}\right)^3 = \left(\frac{72.8 \text{ dynes/cm}}{26.8 \text{ dynes/cm}}\right)^3 = 20.0$$

Reflect: Water has a larger surface tension so the water drop is larger.

13.45. Set Up: The continuity equation is $A_1v_1 = A_2v_2$. $A = \frac{1}{2}\pi d^2$, where d is the pipe diameter.

Solve: $\frac{1}{2}\pi d_1^2 v_1 = \frac{1}{2}\pi d_2^2 v_2$ and $v_2 = \left(\frac{d_1}{d_2}\right)^2 v_1 = \left(\frac{2.50 \text{ in.}}{1.00 \text{ in.}}\right)^2 (6.00 \text{ cm/s}) = 37.5 \text{ cm/s}$

Reflect: To achieve the same volume flow rate the water flows faster in the smaller diameter pipe. Note that the pipe diameters entered in a ratio so there was no need to convert units.

13.49. Set Up: Let point 1 be in the mains and point 2 be in the emerging stream at the end of the fire hose. $v_1 \approx 0$. $y_1 - y_2 = 0$. $p_2 = p_{air}$. After the water emerges from the hose, with upward velocity, it moves in free fall.

Solve: $p_1 + \rho g y_1 + \frac{1}{2}\rho v_1^2 = p_2 + \rho g y_2 + \frac{1}{2}\rho v_2^2$ gives $p_1 - p_{air} = \frac{1}{2}\rho v_2^2$. The motion of a drop of water after it leaves the hose gives $\frac{1}{2}mv_2^2 = mgh$.

$$v_2^2 = 2gh = 2(9.80 \text{ m/s}^2)(15.0 \text{ m}) = 294 \text{ m}^2/\text{s}^2.$$

$$p_1 - p_{air} = \frac{1}{2}(1000 \text{ kg/m}^3)(294 \text{ m}^2/\text{s}^2) = 1.47 \times 10^5 \text{ Pa}$$

Reflect: This is the gauge pressure required to support a column of water 15.0 m high.

13.51. Set Up: Let point 1 be at the top surface and point 2 be at the bottom surface. Neglect the $\rho g(y_1 - y_2)$ term in Bernoulli's equation. In calculating the net force, ΣF_y, take $+y$ to be upward.

Solve: $p_1 + \rho g y_1 + \frac{1}{2}\rho v_1^2 = p_2 + \rho g y_2 + \frac{1}{2}\rho v_2^2$.

$$p_2 - p_1 = \frac{1}{2}\rho(v_1^2 - v_2^2) = \frac{1}{2}(1.20 \text{ kg/m}^3)([70.0 \text{ m/s}]^2 - [60.0 \text{ m/s}]^2) = 780 \text{ Pa}$$

$$\Sigma F_y = p_2A - p_1A - mg = (780 \text{ Pa})(16.2 \text{ m}^2) - (1340 \text{ kg})(9.80 \text{ m/s}^2) = -500 \text{ N}.$$

The net force is 500 N, downward.

Reflect: The pressure is lower where the fluid speed is higher.

13.55. Set Up: $y_1 = y_2$. $v_1A_1 = v_2A_2$. $A_2 = 2A_1$.

Solve: $p_1 + \rho g y_1 + \frac{1}{2}\rho v_1^2 = p_2 + \rho g y_2 + \frac{1}{2}\rho v_2^2$.

$$v_2 = v_1\left(\frac{A_1}{A_2}\right) = (2.50 \text{ m/s})\left(\frac{A_1}{2A_1}\right) = 1.25 \text{ m/s}.$$

$$p_2 = p_1 + \frac{1}{2}\rho(v_1^2 - v_2^2) = 1.80 \times 10^4 \text{ Pa} + \frac{1}{2}(1000 \text{ kg/m}^3)([2.50 \text{ m/s}]^2 - [1.25 \text{ m/s}]^2) = 2.03 \times 10^4 \text{ Pa}$$

Reflect: The gauge pressure is higher at the second point because the water speed is less there.

13.59. Set Up: The flow rate, $\Delta V/\Delta t$, is related to the radius R or diameter D of the artery by Poiseuille's law:

$$\frac{\Delta V}{\Delta t} = \frac{\pi R^4}{8\eta}\left(\frac{p_1 - p_2}{L}\right) = \frac{\pi D^4}{128\eta}\left(\frac{p_1 - p_2}{L}\right).$$

Assume the pressure gradient $(p_1 - p_2)/L$ in the artery remains the same.

Solve: $(\Delta V/\Delta t)/D^4 = \dfrac{\pi}{128\eta}\left(\dfrac{p_1 - p_2}{L}\right) = \text{constant},$ so $(\Delta V/\Delta t)_{\text{old}}/D_{\text{old}}^4 = (\Delta V/\Delta t)_{\text{new}}/D_{\text{new}}^4.$ $(\Delta V/\Delta t)_{\text{new}} = 2(\Delta V/\Delta t)_{\text{old}}$ and $D_{\text{old}} = D.$ This gives

$$D_{\text{new}} = D_{\text{old}}\left[\frac{(\Delta V/\Delta t)_{\text{new}}}{(\Delta V/\Delta t)_{\text{old}}}\right]^{1/4} = 2^{1/4}D = 1.19D.$$

Reflect: Since the flow rate is proportional to D^4, a 19% increase in D doubles the flow rate.

13.61. Set Up: $\rho_{\text{Hg}} = 13.6 \times 10^3 \text{ kg/m}^3$. Let x be the height of the mercury surface in the right arm above the level of the mercury-water interface in the left arm.
Solve: (a) $p - p_{\text{air}} = \rho_{\text{water}}gh = (1.00 \times 10^3 \text{ kg/m}^3)(9.80 \text{ m/s}^2)(0.150 \text{ m}) = 1.47 \times 10^3 \text{ Pa}$
(b) The gauge pressure a distance x below the mercury surface in the right arm equals the gauge pressure at the mercury-water interface, since these two points are at the same height in the mercury. $\rho_{\text{Hg}}gx = 1.47 \times 10^3 \text{ Pa}$ and

$$x = \frac{1.47 \times 10^3 \text{ Pa}}{(13.6 \times 10^3 \text{ kg/m}^3)(9.80 \text{ m/s}^2)} = 1.1 \text{ cm}.$$

$h + x = 15.0 \text{ cm}$, so $h = 13.9 \text{ cm}$.
Reflect: The weight of the water pushes the mercury down in the left-hand arm. A 1.1 cm column of mercury produces the same pressure as a 15.0 cm column of water.

13.63. Set Up: The density of lead is $11.3 \times 10^3 \text{ kg/m}^3$. The buoyant force when the wood sinks is $F_B = \rho_{\text{water}}V_{\text{tot}}g$, where V_{tot} is the volume of the wood plus the volume of the lead. $\rho = m/V$.
Solve: $V_{\text{wood}} = (0.600 \text{ m})(0.250 \text{ m})(0.080 \text{ m}) = 0.0120 \text{ m}^3$.

$$m_{\text{wood}} = \rho_{\text{wood}}V_{\text{wood}} = (600 \text{ kg/m}^3)(0.0120 \text{ m}^3) = 7.20 \text{ kg}.$$

$F_B = (m_{\text{wood}} + m_{\text{lead}})g$. Using $F_B = \rho_{\text{water}}V_{\text{tot}}g$ and $V_{\text{tot}} = V_{\text{wood}} + V_{\text{lead}}$ gives

$$\rho_{\text{water}}(V_{\text{wood}} + V_{\text{lead}})g = (m_{\text{wood}} + m_{\text{lead}})g.$$

$m_{\text{lead}} = \rho_{\text{lead}}V_{\text{lead}}$ then gives $\rho_{\text{water}}V_{\text{wood}} + \rho_{\text{water}}V_{\text{lead}} = m_{\text{wood}} + \rho_{\text{lead}}V_{\text{lead}}$.

$$V_{\text{lead}} = \frac{\rho_{\text{water}}V_{\text{wood}} - m_{\text{wood}}}{\rho_{\text{lead}} - \rho_{\text{water}}} = \frac{(1000 \text{ kg/m}^3)(0.0120 \text{ m}^3) - 7.20 \text{ kg}}{11.3 \times 10^3 \text{ kg/m}^3 - 1000 \text{ kg/m}^3} = 4.66 \times 10^{-4} \text{ m}^3.$$

$m_{\text{lead}} = \rho_{\text{lead}}V_{\text{lead}} = 5.27 \text{ kg}$.
Reflect: The volume of the lead is only 3.9% of the volume of the wood. If the contribution of the volume of the lead to F_B is neglected, the calculation is simplified: $\rho_{\text{water}}V_{\text{wood}}g = (m_{\text{wood}} + m_{\text{lead}})g$ and $m_{\text{lead}} = 4.8 \text{ kg}$. The result of this calculation is in error by about 9%.

13.67. Set Up: The density of gold is $\rho_g = 19.3 \times 10^3 \text{ kg/m}^3$ and the density of aluminum is $\rho_{\text{al}} = 2.7 \times 10^3 \text{ kg/m}^3$. $F_B = \rho_w V_{\text{tot}}g$.
Solve: Find the volume V_{tot} of the ingot: $F_B + T - w = 0$. $\rho_w V_{\text{tot}}g = W - T$.

$$V_{\text{tot}} = \frac{W - T}{\rho_w g} = \frac{45.0 \text{ N} - 39.0 \text{ N}}{(1000 \text{ kg/m}^3)(9.80 \text{ m/s}^2)} = 6.12 \times 10^{-4} \text{ m}^3.$$

The mass of the ingot is $m_{tot} = W/g = 4.59$ kg. The total mass is the mass of gold, m_g, plus the mass of aluminum, m_{al}. The total volume is the volume of gold, V_g, plus the volume of aluminum, V_{al}.

$$m_{tot} = m_g + m_{al} = \rho_g V_g + \rho_{al} V_{al} = \rho_g V_g + \rho_{al}(V_{tot} - V_g).$$

$$V_g = \frac{m_{tot} - \rho_{al} V_{tot}}{\rho_g - \rho_{al}} = \frac{4.59 \text{ kg} - (2700 \text{ kg/m}^3)(6.12 \times 10^{-4} \text{ m}^3)}{19.3 \times 10^3 \text{ kg/m}^3 - 2.7 \times 10^3 \text{ kg/m}^3} = 1.77 \times 10^{-4} \text{ m}^3.$$

$$w_g = m_g g = \rho_g V_g g = (19.3 \times 10^3 \text{ kg/m}^3)(1.77 \times 10^{-4} \text{ m}^3)(9.80 \text{ m/s}^2) = 33.5 \text{ N}.$$

Reflect: If the ingot had the same weight in air (45.0 N) and was pure gold, its volume would be $2.38 \times 10^{-4} \text{ m}^3$ and the balance would read 42.7 N when the ingot was submerged.

13.69. Set Up: Apply $p_1 + \rho g y_1 + \frac{1}{2}\rho v_1^2 = p_2 + \rho g y_2 + \frac{1}{2}\rho v_2^2$, with point 1 at the surface of the acid in the tank and point 2 in the stream as it emerges from the hole. $p_1 = p_2 = p_{air}$. Since the hole is small the level in the tank drops slowly and $v_1 \approx 0$. After a drop of acid exits the hole the only force on it is gravity and it moves in projectile motion. For the projectile motion take $+y$ downward, so $a_x = 0$ and $a_y = +9.80 \text{ m/s}^2$.
Solve: Bernoulli's equation with $p_1 = p_2$ and $v_1 = 0$ gives

$$v_2 = \sqrt{2g(y_1 - y_2)} = \sqrt{2(9.80 \text{ m/s}^2)(0.75 \text{ m})} = 3.83 \text{ m/s}.$$

projectile motion: Use the vertical motion to find the time in the air. $v_{0y} = 0$, $a_y = +9.80 \text{ m/s}^2$, $y - y_0 = 1.4$ m. $y - y_0 = v_{0y}t + \frac{1}{2}a_y t^2$ gives

$$t = \sqrt{\frac{2(y - y_0)}{a_y}} = \sqrt{\frac{2(1.4 \text{ m})}{9.80 \text{ m/s}^2}} = 0.535 \text{ s}$$

The horizontal distance a drop travels in this time is $x - x_0 = v_{0x}t + \frac{1}{2}a_x t^2 = (3.83 \text{ m/s})(0.535 \text{ s}) = 2.05 \text{ m}$.
Reflect: If the depth of acid in the tank is increased, then the velocity of the stream as it emerges from the hole increases and the horizontal range of the stream increases.

13.71. Set Up: For a spherical astronomical object, $g = G\dfrac{m}{R^2}$. $p = p_{air} + \rho g h$.
Solve: (a) Find g on Europa:

$$g = (6.673 \times 10^{-11} \text{ N} \cdot \text{m}^2/\text{kg}^2)\frac{4.78 \times 10^{22} \text{ kg}}{(1.565 \times 10^6 \text{ m})^2} = 1.30 \text{ m/s}^2$$

The gauge pressure at a depth of 100 m would be

$$p - p_{air} = \rho g h = (1.00 \times 10^3 \text{ kg/m}^3)(1.30 \text{ m/s}^2)(100 \text{ m}) = 1.30 \times 10^5 \text{ Pa}$$

(b) Solve for h on earth that gives $p - p_{air} = 1.30 \times 10^5$ Pa:

$$h = \frac{p - p_{air}}{\rho g} = \frac{1.30 \times 10^5 \text{ Pa}}{(1.00 \times 10^3 \text{ kg/m}^3)(9.80 \text{ m/s}^2)} = 13.3 \text{ m}$$

Reflect: g on Europa is less than on earth so the pressure at a water depth of 100 m is much less on Europa than it would be on earth.

TEMPERATURE AND HEAT

Problems 5, 7, 13, 15, 17, 21, 23, 27, 29, 33, 35, 37, 39, 43, 45, 47, 53, 55, 57, 59, 61, 65, 67, 69, 71, 75, 81, 83, 85, 87, 91

Solutions to Problems

14.5. Set Up and Solve: (a) $T_F = \frac{9}{5}T_C + 32$. Set $T_F = T_C = T$. $T = \frac{9}{5}T + 32$. $\frac{4}{5}T = -32$ and $T = -40°$. $-40°C = -40°F$.
(b) $T_K = T_C + 273.15$. It is not possible to have $T_K = T_C$.
Reflect: Any Celsius temperature is less than the corresponding Kelvin temperature.

14.7. Set Up: (a) $T_0 = -8.0°C$ and $T = 40.0°C$. $L_0 = 984$ ft and we are asked to solve for ΔL. From Table 14.1, the coefficient of linear expansion is $1.2 \times 10^{-5} \, (C°)^{-1}$. **(b)** Use $9 \, F° = 5 \, C°$.
Solve: (a) $\Delta L = \alpha L_0 \, \Delta T = [1.2 \times 10^{-5} \, (C°)^{-1}][984 \text{ ft}][40.0°C - (-8.0°C)] = 0.57$ ft $= 6.8$ in.
(b) $\alpha = [1.2 \times 10^{-5} \, (C°)^{-1}][5 \, C°/9 \, F°] = 6.7 \times 10^{-6} \, (F°)^{-1}$
Reflect: In (a) note that the fractional change in length, $\Delta L/L_0$, is very small. In (b), α is smaller in $(F°)^{-1}$ than in $(C°)^{-1}$. A Fahrenheit degree is smaller than a Celsius degree and a temperature difference is a larger number when expressed in F° than in C°. So, to give the same $\alpha \, \Delta T$ value, α is smaller in $(F°)^{-1}$.

14.13. Set Up: The capacity of a flask contracts when the glass is cooled, with the same volume change as for a solid piece of glass. $\beta = 3\alpha$.
Solve: $\beta = 3\alpha = 3.60 \times 10^{-5} \, (C°)^{-1}$.

$$V = V_0(1 + \beta \, \Delta T) = (500.00 \text{ mL})(1 + [3.60 \times 10^{-5} \, (C°)^{-1}][-15.0 \, C°]) = 499.73 \text{ mL}.$$

Reflect: The fractional change in the volume is very small.

14.15. Set Up: For mercury, $\beta_{Hg} = 18 \times 10^{-5} \, (C°)^{-1}$. When heated, both the volume of the flask and the volume of the mercury increase. 8.95 cm^3 of mercury overflows, so $\Delta V_{Hg} - \Delta V_{glass} = 8.95$ cm^3.
Solve: $\Delta V_{Hg} = V_0 \beta_{Hg} \, \Delta T = (1000.00 \text{ cm}^3)(18 \times 10^{-5} \, (C°)^{-1})(55.0 \, C°) = 9.9$ cm^3.

$$\Delta V_{glass} = \Delta V_{Hg} - 8.95 \text{ cm}^3 = 0.95 \text{ cm}^3.$$

$$\beta_{glass} = \frac{\Delta V_{glass}}{V_0 \, \Delta T} = \frac{0.95 \text{ cm}^3}{(1000.00 \text{ cm}^3)(55.0 \, C°)} = 1.7 \times 10^{-5} \, (C°)^{-1}.$$

Reflect: The coefficient of volume expansion for the mercury is larger than for glass. When they are heated, both the volume of the mercury and the inside volume of the flask increase. But the increase for the mercury is greater and it no longer all fits inside the flask.

14.17. Set Up: $L_0 = 20.0$ cm at $0.00°C$. For aluminum, $\alpha_{al} = 2.4 \times 10^{-5}$ $(C°)^{-1}$. For brass, $\alpha_{br} = 2.0 \times 10^{-5}$ $(C°)^{-1}$. The distance between the marks at $100.0°C$ is $L_{al} - L_{br}$.
Solve: $L_{al} = L_0(1 + \alpha_{al} \Delta T)$. $L_{br} = L_0(1 + \alpha_{br} \Delta T)$.

$$L_{al} - L_{br} = L_0(\alpha_{al} - \alpha_{br}) \Delta T = (20.0 \text{ cm})(0.4 \times 10^{-5} (C°)^{-1})(100.0 \text{ C°}) = 8.0 \times 10^{-3} \text{ cm}.$$

Reflect: This section of both rulers increases in length, but the aluminum one increases more since its coefficient of linear expansion is greater.

14.21. Set Up: 0.50 L of air has mass 0.65×10^{-3} kg.
Solve: (a) $Q = mc \Delta T = (0.65 \times 10^{-3} \text{ kg})(1020 \text{ J/kg} \cdot \text{K})(57 \text{ C°}) = 38$ J
(b) $Q = (38 \text{ J/breath})(20 \text{ breaths/min})(60 \text{ min/h}) = 4.6 \times 10^4$ J/h.
Reflect: Air has a small density, so one liter of air has little mass. But the specific heat capacity of air is rather large, larger than for most metals and about one-fourth the value for water.

14.23. Set Up: (a) The initial kinetic energy $K = \frac{1}{2}mv^2$ of the car is converted to heat. **(b)** The heat Q calculated in (a) causes a temperature rise of the car. From Table 14.3, the specific heat capacity of iron is $c = 0.47 \times 10^3$ J/kg $\cdot$ C°
Solve: (a) The initial kinetic energy of the car is $K_i = \frac{1}{2}mv_i^2 = \frac{1}{2}(1525 \text{ kg})(25.0 \text{ m/s})^2 = 4.77 \times 10^5$ J. The car stops, so $K_f = 0$. $Q = |\Delta K| = 4.77 \times 10^5$ J.
(b) $Q = mc \Delta T$ so

$$\Delta T = \frac{Q}{mc} = \frac{4.77 \times 10^5 \text{ J}}{(1525 \text{ kg})(0.47 \times 10^3 \text{ J/kg} \cdot \text{C°})} = 0.666 \text{ C°}$$

Reflect: The stopping time (17 s) did not enter into the calculation.

14.27. Set Up: Set the loss of kinetic energy of the bullet equal to the heat energy Q transferred to the water. From Table 14.3, the specific heat capacity of water is 4.19×10^3 J/kg $\cdot$ C°.
Solve: The kinetic energy lost by the bullet is

$$K_i - K_f = \frac{1}{2}m(v_i^2 - v_f^2) = \frac{1}{2}(15.0 \times 10^{-3} \text{ kg})([865 \text{ m/s}]^2 - [534 \text{ m/s}]^2) = 3.47 \times 10^3 \text{ J},$$

so for the water $Q = 3.47 \times 10^3$ J. $Q = mc \Delta T$ gives

$$\Delta T = \frac{Q}{mc} = \frac{3.47 \times 10^3 \text{ J}}{(13.5 \text{ kg})(4.19 \times 10^3 \text{ J/kg} \cdot \text{C°})} = 0.0613 \text{ C°}$$

Reflect: The heat energy required to change the temperature of ordinary size objects is very large compared to the typical kinetic energies of moving objects.

14.29. Set Up: $1 \text{ W} = 1 \text{ J/s}$
Solve: (a) $Q = (65.0 \text{ W})(120 \text{ s}) = 7.8 \times 10^3$ J. $Q = mc \Delta T$ and

$$c = \frac{Q}{m \Delta T} = \frac{7.8 \times 10^3 \text{ J}}{(0.780 \text{ kg})(3.99 \text{ C°})} = 2.51 \times 10^3 \text{ J/kg} \cdot \text{K}.$$

(b) If some of the input energy doesn't go into the liquid the Q for the calculation of c should be smaller and the correct c would be smaller than the value we calculated in (a), The result in (a) is an overestimate.
Reflect: The specific heat capacity of the liquid is about 60% of the value for water.

14.33. Set Up: The heat that comes out of the person goes into the ice-water bath and causes some of the ice to melt. Normal body temperature is $98.6°F = 37.0°C$, so for the person $\Delta T = -5$ C°. The ice-water bath stays at $0°C$. A mass m of ice melts and $Q_{ice} = mL_f$. From Table 14.4, for water $L_f = 334 \times 10^3$ J/kg.
Solve: $Q_{person} = mc \Delta T = (70.0 \text{ kg})(3480 \text{ J/kg} \cdot \text{C°})(-5.0 \text{ C°}) = -1.22 \times 10^6$ J
Therefore, the amount of that that goes into the ice is 1.22×10^6 J. $m_{ice}L_f = 1.22 \times 10^6$ J and

$$m_{ice} = \frac{1.22 \times 10^6 \text{ J}}{334 \times 10^3 \text{ J/kg}} = 3.7 \text{ kg}.$$

Reflect: If less ice than this is used, all the ice melts and the temperature of the water in the bath rises above $0°C$.

14.35. Set Up: The initial mechanical energy of the bullet is $K = \frac{1}{2}mv^2$. For lead, $c = 0.13 \times 10^3$ J/kg $\cdot$ K, the melting point is 327.3°C and $L_f = 2.45 \times 10^4$ J/kg.
Solve: $K = Q$. $\frac{1}{2}mv^2 = mc \, \Delta T + mL_f$, where $\Delta T = 327.3°C - 25°C = 302$ C°.

$$v = \sqrt{2(c \, \Delta T + L_f)} = \sqrt{2([0.13 \times 10^3 \text{ J/kg} \cdot \text{K}][302 \text{ C}°] + 2.45 \times 10^4 \text{ J/kg})} = 357 \text{ m/s}.$$

Reflect: The answer is independent of the mass of the bullet.

14.37. Set Up: For water, $c = 4.19 \times 10^3$ J/kg $\cdot$ K and $L_f = 3.34 \times 10^5$ J/kg.
Solve: $Q = mc \, \Delta T - mL_f = (0.350 \text{ kg})([4.19 \times 10^3 \text{ J/kg} \cdot \text{K}][-18.0 \text{ C}°] - 3.34 \times 10^5 \text{ J/kg}) = -1.43 \times 10^5$ J. The minus sign says 1.43×10^5 J must be removed from the water.

$$(1.43 \times 10^5 \text{ J})\left(\frac{1 \text{ cal}}{4.186 \text{ J}}\right) = 3.42 \times 10^4 \text{ cal} = 34.2 \text{ kcal}.$$

Reflect: $Q < 0$ when heat comes out of an object the equation $Q = mc \, \Delta T$ puts in the correct sign automatically, from the sign of $\Delta T = T_f - T_i$. But in $Q = \pm mL$ we must select the correct sign.

14.39. Set Up: For water, $L_v = 2.256 \times 10^6$ J/kg and $c = 4.19 \times 10^3$ J/kg $\cdot$ K.
Solve: (a) $Q = +m(-L_v + c \, \Delta T)$

$$Q = +(25.0 \times 10^{-3} \text{ kg})(-2.256 \times 10^6 \text{ J/kg} + [4.19 \times 10^3 \text{ J/kg} \cdot \text{K}][-66.0 \text{ C}°]) = -6.33 \times 10^4 \text{ J}$$

(b) $Q = mc \, \Delta T = (25.0 \times 10^{-3} \text{ kg})(4.19 \times 10^3 \text{ J/kg} \cdot \text{K})(-66.0 \text{ C}°) = -6.91 \times 10^3$ J.
(c) The total heat released by the water that starts as steam is nearly a factor of ten larger than the heat released by water that starts at 100°C. Steam burns are much more severe than hot-water burns.
Reflect: For a given amount of material, the heat for a phase change is typically much more than the heat for a temperature change.

14.43. Set Up: For water, $c = 4.19 \times 10^3$ J/kg $\cdot$ K.
Solve: (a) $Q_{\text{water}} = mc \, \Delta T = (1.00 \text{ kg})(4.19 \times 10^3 \text{ J/kg} \cdot \text{K})(2.0 \text{ C}°) = +8.38 \times 10^3$ J. $Q_{\text{metal}} = mc \, \Delta T = (0.500 \text{ kg})c(-78 \text{ C}°)$. $\Sigma Q = 0$ says $(0.500 \text{ kg})c(-78 \text{ C}°) + 8.38 \times 10^3 \text{ J} = 0$.

$$c = \frac{8.38 \times 10^3 \text{ J}}{(0.500 \text{ kg})(78 \text{ C}°)} = 215 \text{ J/kg} \cdot \text{K}.$$

(b) Water has a much larger value of c so stores more heat for the same ΔT.
(c) If some of the heat went into the Styrofoam™ then more would have to come out of the metal and c would be found to be greater. The value calculated in (a) would be too small.
Reflect: The amount of heat that comes out of the metal when it cools equals the amount of heat that goes into the water to produce its temperature increase.

14.45. Set Up: The total mass of the coins is $m_c = 0.600$ kg. For water, $c_w = 4.19 \times 10^3$ J/kg $\cdot$ K.
Solve: For the coins,

$$Q_c = m_c c_c \, \Delta T_c = (0.600 \text{ kg})(390 \text{ J/kg} \cdot \text{K})(T - 100.0°C) = (234 \text{ J/K})T - 2.34 \times 10^4 \text{ J}.$$

For the water,

$$Q_w = m_w c_w \, \Delta T_w = (0.240 \text{ kg})(4190 \text{ J/kg} \cdot \text{K})(T - 20.0°C) = (1006 \text{ J/K})T - 2.01 \times 10^4 \text{ J}.$$

$\Sigma Q = 0$ gives $(234 \text{ J/K})T - 2.34 \times 10^4 \text{ J} + (1006 \text{ J/K})T - 2.01 \times 10^4 \text{ J} = 0$.

$$T = \frac{4.35 \times 10^4 \text{ J}}{1240 \text{ J/K}} = 35.1°C.$$

Reflect: We assumed that the water remained liquid. This is a correct assumption since the final temperature we calculated is in the range $0°C < T < 100°C$.

14.47. Set Up: Since some ice remains, the ice and water from the melted ice remain at 0°C. For silver, $c_s = 230$ J/kg · K. For water, $L_f = 3.34 \times 10^5$ J/kg.

Solve: For the silver, $Q_s = m_s c_s \Delta T_s = (4.00 \text{ kg})(230 \text{ J/kg} \cdot \text{K})(-750 \text{ C}°) = -6.90 \times 10^5$ J.

For the ice, $Q_i = +mL_f$, where m is the mass that melts.

$\Sigma Q = 0$ gives $m(3.34 \times 10^5 \text{ J/kg}) - 6.90 \times 10^5 \text{ J} = 0$ and $m = 2.07$ kg.

Reflect: The heat that comes out of the ingot when it cools goes into the ice to produce a phase change.

14.53. Set Up: Apply $H = kA\dfrac{T_H - T_C}{L}$ and solve for k.

Solve: $k = \dfrac{HL}{A(T_H - T_C)} = \dfrac{(75 \text{ W})(0.75 \times 10^{-3} \text{ m})}{(2.0 \text{ m}^2)(37°\text{C} - 30.0°\text{C})} = 4.0 \times 10^{-3}$ W/m · C°.

Reflect: This is a small value; skin is a poor conductor of heat. But the thickness of the skin is small, so the rate of heat conduction through the skin is not small.

14.55. Set Up: The heat current Q/t is the same through the wood as through the Styrofoam™.

Solve: (a) $\dfrac{Q}{t} = \dfrac{kA(T_H - T_C)}{L}$ and $\left(\dfrac{Q}{t}\right)_w = \left(\dfrac{Q}{t}\right)_s$ gives $\dfrac{k_w A(T - [-10.0°\text{C}])}{L_w} = \dfrac{k_s A(19.0°\text{C} - T)}{L_s}$.

$$\dfrac{(0.080 \text{ W/m} \cdot \text{K})(T + 10.0°\text{C})}{0.030 \text{ m}} = \dfrac{(0.010 \text{ W/m} \cdot \text{K})(19.0°\text{C} - T)}{0.022 \text{ m}}.$$

$2.67(T + 10.0°\text{C}) = 0.455(19.0°\text{C} - T)$ and $T = \dfrac{26.7°\text{C} - 8.65°\text{C}}{-3.125} = -5.8°\text{C}.$

(b) $\left(\dfrac{Q}{tA}\right)_w = \dfrac{k(T_H - T_C)}{L} = \dfrac{(0.080 \text{ W/m} \cdot \text{K})(-5.8°\text{C} + 10.0°\text{C})}{0.030 \text{ m}} = 11$ W/m²

Or, $\left(\dfrac{Q}{tA}\right)_s = \dfrac{(0.010 \text{ W/m} \cdot \text{K})(19.0°\text{C} - [-5.8°\text{C}])}{0.022 \text{ m}} = 11$ W/m², which checks.

Reflect: k is much smaller for the Styrofoam™ so the temperature gradient across it is much larger than across the wood.

14.57. Set Up: For water, $L_f = 3.34 \times 10^5$ J/kg.

Solve: The heat conducted by the rod in 10.0 min is

$$Q = mL_f = (8.50 \times 10^{-3} \text{ kg})(3.34 \times 10^5 \text{ J/kg}) = 2.84 \times 10^3 \text{ J}.$$

$$\dfrac{Q}{t} = \dfrac{2.84 \times 10^3 \text{ J}}{600 \text{ s}} = 4.73 \text{ W}.$$

$$\dfrac{Q}{t} = \dfrac{kA(T_H - T_C)}{L}.$$

$$k = \dfrac{(Q/t)L}{A(T_H - T_C)} = \dfrac{(4.73 \text{ W})(0.600 \text{ m})}{(1.25 \times 10^{-4} \text{ m}^2)(100 \text{ C}°)} = 227 \text{ W/m} \cdot \text{K}.$$

Reflect: The heat conducted by the rod is the heat that enters the ice and produces the phase change.

14.59. Set Up: Let the temperature of the fat-air boundary be T. A section of the two layers is sketched in Figure 14.59. A Kelvin degree is the same size as a Celsius degree, so $W/m \cdot K$ and $W/m \cdot C°$ are equivalent units. At steady state the heat current through each layer is the same, equal to 50 W. The area of each layer is $A = 4\pi r^2$, with $r = 0.75$ m.

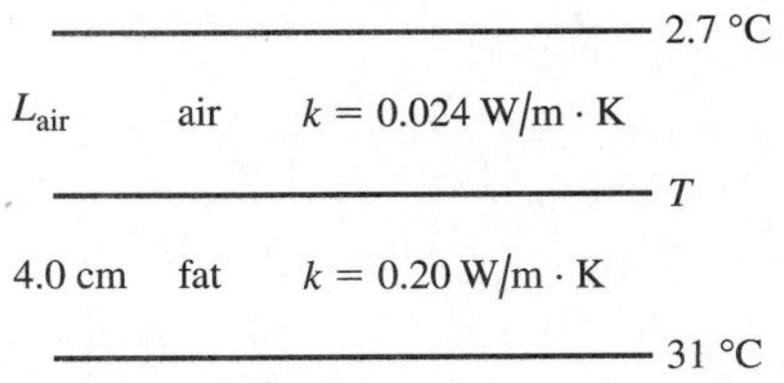

Figure 14.59

Solve: (a) Apply $H = kA\dfrac{T_H - T_C}{L}$ to the fat layer and solve for $T_C = T$. For the fat layer $T_H = 31°C$.

$$T = T_H - \frac{HL}{kA} = 31°C - \frac{(50\ \text{W})(4.0 \times 10^{-2}\ \text{m})}{(0.20\ \text{W})(4\pi)(0.75\ \text{m})^2} = 31°C - 1.4°C = 29.6°C$$

(b) Apply $H = kA\dfrac{T_H - T_C}{L}$ to the air layer and solve for $L = L_{air}$. For the air layer $T_H = T = 29.6°C$ and $T_C = 2.7°C$.

$$L = \frac{kA(T_H - T_C)}{H} = \frac{(0.024\ \text{W})(4\pi)(0.75\ \text{m})^2(29.6°C - 2.7°C)}{50\ \text{W}} = 9.1\ \text{cm}$$

Reflect: The thermal conductivity of air is much lass than the thermal conductivity of fat, so the temperature gradient for the air must be much larger to achieve the same heat current. So, most of the temperature difference is across the air layer.

14.61. Set Up: (a) $M = 75$ and $h = 1.83$. **(b)** and **(c)** $e = 1.00$. In $H = Ae\sigma T^4$, T must be in kelvins. $T = 30°C = 303$ K. $T_s = 20°C = 293$ K.
Solve: (a) $A = (0.202)(75)^{0.425}(1.83)^{0.725} = 1.96\ \text{m}^2$
(b) $H = Ae\sigma T^4 = (1.96\ \text{m}^2)(1.00)(5.67 \times 10^{-8}\ \text{W}/\text{m}^2 \cdot \text{K}^4)(303)^4 = 937\ \text{W}$
(c) $H_{net} = Ae\sigma(T^4 - T_s^4)$

$$H_{net} = (1.96\ \text{m}^2)(1.00)(5.67 \times 10^{-8}\ \text{W}/\text{m}^2 \cdot \text{K}^4)([303\ \text{K}]^4 - [293\ \text{K}]^4) = 118\ \text{W}$$

(d) The surface area of a sphere is $4\pi r^2$; use this for your head. Treat each arm, each leg and your trunk as cylinders; the surface area of a cylinder of radius r and length l is $2\pi r^2 + 2\pi rl$. Should you include the πr^2 area of each end of the cylinders?
Reflect: Since the radiation heat current is proportional to T^4, the net rate of heat loss increases rapidly when the air temperature decreases.

14.65. Set Up and Solve: $H = Ae\sigma T^4$.

$$A = \frac{H}{e\sigma T^4} = \frac{150\ \text{W}}{(0.35)(5.67 \times 10^{-8}\ \text{W}/\text{m}^2 \cdot \text{K}^4)(2450\ \text{K})^4} = 2.10\ \text{cm}^2.$$

Reflect: Even though the filament is small, its coil shape gives it a relatively large surface area.

14.67. Set Up: The heat energy generated by friction work equals the loss of kinetic energy. The heat Q_{ice} that goes into the ice is related to the mass m_{melts} of ice that melts by $Q_{ice} = m_{melts}L_f$. For ice, $L_f = 334 \times 10^3\ \text{J/kg}$. The maximum amount of ice melts when the block loses all its kinetic energy and comes to rest.
Solve: (a) For $v_i = 15.0\ \text{m/s}$ and $v_f = 10.0\ \text{m/s}$,

$$Q = K_i - K_f = \tfrac{1}{2}m(v_i^2 - v_f^2) = \tfrac{1}{2}(8.5\ \text{kg})([15.0\ \text{m/s}]^2 - [10.0\ \text{m/s}]^2) = 531\ \text{J}.$$

$Q_{ice} = \frac{1}{2}Q = 265$ J. Then $Q_{ice} = mL_f$ gives

$$m = \frac{Q_{ice}}{L_f} = \frac{265 \text{ J}}{334 \times 10^3 \text{ J/kg}} = 7.93 \times 10^{-4} \text{ kg} = 0.793 \text{ g}$$

(b) For $v_f = 0$, $Q = K_i = 956$ J. $Q_{ice} = 478$ J and $m = \dfrac{Q_{ice}}{L_f} = 1.43$ g

Reflect: Since the ice is at 0°C, any heat that goes into it causes a phase transition. 15.0 m/s is 34 mph. When brought to rest from this large initial speed only a very small fraction of the ice melts. Also note the irreversible nature of this; the water won't refreeze and release energy to get the block moving again.

14.69. Set Up: Write the volume of the oceans as $V = Ad$, where d is the average depth and A is the average surface area of the oceans. Assume that only d changes when V changes, so $\Delta V = A\Delta d$.
Solve: $\Delta V = V_0 \beta \, \Delta T$, $\Delta V = A\Delta d$ and $V_0 = Ad_0$, with $d_0 = 4000$ m. This gives

$$\Delta d = d_0 \beta \, \Delta T = (4000 \text{ m})(0.207 \times 10^{-3} \text{ (C°)}^{-1})(3.5 \text{ C°}) = 2.9 \text{ m}.$$

Reflect: This rise in sea level would have serious effects on coastal areas.

14.71. Set Up: 1.00 L of water has a mass of 1.00 kg, so

$$9.46 \text{ L/min} = (9.46 \text{ L/min})(1.00 \text{ kg/L})(1 \text{ min/60 s}) = 0.158 \text{ kg/s}.$$

For water, $c = 4190$ J/kg·C°.
Solve: $Q = mc \, \Delta T$ so

$$H = (Q/t) = (m/t)c \, \Delta T = (0.158 \text{ kg/s})(4190 \text{ J/kg·C°})(49°C - 10°C) = 2.6 \times 10^4 \text{ W} = 26 \text{ kW}$$

Reflect: The power requirement is large, the equivalent of 260 light bulbs that are 100-W, but this large power is needed only for short periods of time.

14.75. Set Up: Let the cross sectional area of the cup be A. For the cup, $V_{0,cup} = A(10.0 \text{ cm})$. For the oil, $V_{0,oil} = A(9.9 \text{ cm})$. Oil starts to overflow when $\Delta V_{oil} = \Delta V_{cup} + (0.100 \text{ cm})A$.
Solve: $\Delta V_{cup} = V_{0,cup}\beta_{glass} \, \Delta T$. $\Delta V_{oil} = V_{0,oil}\beta_{oil} \, \Delta T$. $(9.9 \text{ cm})A\beta_{oil} \, \Delta T = (10.0 \text{ cm})A\beta_{glass} \, \Delta T + (0.100 \text{ cm})A$.

$$\Delta T = \frac{0.100 \text{ cm}}{(9.9 \text{ cm})\beta_{oil} - (10.0 \text{ cm})\beta_{glass}}$$

$$\Delta T = \frac{0.100 \text{ cm}}{(9.9 \text{ cm})(6.8 \times 10^{-4} \text{ (C°)}^{-1}) - (10.0 \text{ cm})(2.7 \times 10^{-5} \text{ (C°)}^{-1})} = 15.5 \text{ C°}.$$

$T_f = T_i + \Delta T = 37.5$ °C.
Reflect: The olive oil expands more than the capacity of the cup does because it has a much larger coefficient of volume expansion.

14.81. Set Up: For water, $c = 4.19 \times 10^3$ J/kg·K, $L_f = 3.34 \times 10^5$ J/kg, $L_v = 2.256 \times 10^6$ J/kg. If the final temperature is 30.0°C, all the ice melts and all the steam condenses.
Solve:

$$Q_{ice} = m_{ice}L_f + m_{ice}c_{water} \, \Delta T$$

$$Q_{ice} = (0.20 \text{ kg})(3.34 \times 10^5 \text{ J/kg}) + (0.20 \text{ kg})(4.19 \times 10^3 \text{ J/kg·K})(30.0 \text{ C°}) = 9.19 \times 10^4 \text{ J}$$

$$Q_{water} = m_{water}c_{water} \, \Delta T = (2.10 \text{ kg})(4.19 \times 10^3 \text{ J/kg·K})(30.0 \text{ C°}) = 2.64 \times 10^5 \text{ J}$$

$$Q_{steam} = -m_{steam}L_f + m_{steam}c_{water} \, \Delta T$$

$$Q_{steam} = m(-2.256 \times 10^6 \text{ J/kg} + [4.19 \times 10^3 \text{ J/kg·K}][-70.0 \text{ C°}]) = -(2.55 \times 10^6 \text{ J/kg})m$$

$\Sigma Q = 0$, so 9.19×10^4 J $+ 2.64 \times 10^5$ J $- (2.55 \times 10^6 \text{ J/kg})m = 0$ and $m = 0.140$ kg.
Reflect: For water, $L_v > L_f$. A small mass of steam melts a larger mass of ice and also raises the temperature of the water.

14.83. Set Up: The emissivity of a human body is taken to be 1.0. In the equation for the radiation heat current the temperatures must be in kelvins.

Solve: (a) $P_{jog} = (0.80)(1300 \text{ W}) = 1.04 \times 10^3 \text{ J/s}$

(b) $H_{net} = Ae\sigma(T^4 - T_s^4)$

$$H_{net} = (1.85 \text{ m}^2)(1.00)(5.67 \times 10^{-8} \text{ W/m}^2 \cdot \text{K}^4)([306 \text{ K}]^4 - [313 \text{ K}]^4) = -87.1 \text{ W}$$

The person gains 87.1 J of heat each second by radiation.

(c) The total excess heat per second is $1040 \text{ J/s} + 87 \text{ J/s} = 1130 \text{ J/s}$

(d) In 1 min = 60 s the runner must dispose of $(60 \text{ s})(1130 \text{ J/s}) = 6.78 \times 10^4 \text{ J}$. If this much heat goes to evaporate water, the mass m of water that evaporates in one minute is given by $Q = mL_v$, so

$$m = \frac{Q}{L_v} = \frac{6.78 \times 10^4 \text{ J}}{2.42 \times 10^6 \text{ J/kg}} = 0.028 \text{ kg} = 28 \text{ g}.$$

(e) In a half-hour, or 30 minutes, the runner loses $(30 \text{ min})(0.028 \text{ kg/min}) = 0.84 \text{ kg}$. The runner must drink 0.84 L, which is $\dfrac{0.84 \text{ L}}{0.750 \text{ L/bottle}} = 1.1$ bottles.

Reflect: The person gains heat by radiation since the air temperature is greater than his skin temperature.

14.85. Set Up: Assume all the ice melts. If we calculate $T_f < 0$, we will know this assumption is incorrect. For aluminum, $c_a = 910 \text{ J/kg} \cdot \text{K}$. For water, $L_f = 3.35 \times 10^5 \text{ J/kg}$ and $c_w = 4.19 \times 10^3 \text{ J/kg} \cdot \text{K}$. For ice, $c_i = 2.01 \times 10^3 \text{ J/kg} \cdot \text{K}$. The density of water is $1.00 \times 10^3 \text{ kg/m}^3$, so 1.00 L of water has mass 1.00 kg.

Solve: For the soft drink,

$$Q_w = m_w c_w \, \Delta T_w = (2.00 \text{ kg})(4.19 \times 10^3 \text{ J/kg} \cdot \text{K})(T - 20.0°\text{C}) = (8380 \text{ J/K})T - 1.676 \times 10^5 \text{ J}.$$

For the mug, $Q_a = m_a c_a \, \Delta T_a = (0.257 \text{ kg})(910 \text{ J/kg} \cdot \text{K})(T - 20.0°\text{C}) = (234 \text{ J/K})T - 4.68 \times 10^3 \text{ J}$.

For the ice, $Q_i = m_i c_i \, \Delta T_i + m_i L_f + m_i c_w \, \Delta T_w$.

$$Q_i = (0.120 \text{ kg})([2010 \text{ J/kg} \cdot \text{K}][15.0 \text{ C}°] + 3.34 \times 10^5 \text{ J/kg} + [4.19 \times 10^3 \text{ J/kg} \cdot \text{K}][T - 0 \text{ C}°]).$$

$$Q_i = 4.37 \times 10^4 \text{ J} + (503 \text{ J/K})T.$$

$\Sigma Q = 0$ gives

$$(8380 \text{ J/K})T - 1.676 \times 10^5 \text{ J} + (234 \text{ J/K})T - 4.68 \times 10^3 \text{ J} + 4.37 \times 10^4 \text{ J} + (503 \text{ J/K})T = 0.$$

$$T = \frac{1.286 \times 10^5 \text{ J}}{9117 \text{ J/K}} = 14.1°\text{C}.$$

Reflect: $T > 0°\text{C}$, so our assumption that all the ice melts is correct. Note that the ice and the water from the melted ice have different specific heat capacities.

14.87. Set Up: Find the heat Q is conducted into the icebox in 10 hours. The heat from mass m of ice is $Q = mL_f + mc(2.50 \text{ C}°)$.

Solve: $Q = \dfrac{kA(T_H - T_C)t}{L} = \dfrac{(0.050 \text{ W/(m} \cdot \text{K)})(2.00 \text{ m}^2)(15.0 \text{ C}°)(36{,}000 \text{ s})}{3.00 \times 10^{-2} \text{ m}} = 1.80 \times 10^6 \text{ J}$

$$m = \frac{Q}{L_f + c(2.50 \text{ C}°)} = \frac{1.80 \times 10^6 \text{ J}}{3.34 \times 10^5 \text{ J/kg} + (4190 \text{ J/kg} \cdot \text{K})(2.50 \text{ C}°)} = 5.2 \text{ kg}.$$

Reflect: Heat is conducted into the icebox since its outside temperature is greater than its inside temperature. Most of this heat goes into melting the ice and a much smaller amount goes into raising the temperature of the 0°C water that is produced by the phase change.

14.91. Set Up: The radius of the earth is $R_E = 6.38 \times 10^6$ m. Since an ocean depth of 100 m is much less than the radius of the earth, we can calculate the volume of the water to this depth as the surface area of the oceans times 100 m. The total surface area of the earth is $A_{earth} = 4\pi R_E^2$. The density of seawater is 1.03×10^3 kg/m^3 (Table 13.1).

Solve: The surface area of the oceans is $(2/3)A_{earth} = (2/3)(4\pi)(6.38 \times 10^6 \text{ m})^2 = 3.41 \times 10^{14} \text{ m}^2$. The total rate of solar energy incident on the oceans is $(1050 \text{ W/m}^2)(3.41 \times 10^{14} \text{ m}^2) = 3.58 \times 10^{17}$ W. 12 hours per day for 30 days is $(12)(30)(3600)$ s $= 1.30 \times 10^6$ s, so the total solar energy input to the oceans in one month is $(1.30 \times 10^6 \text{ s})(3.58 \times 10^{17} \text{ W}) = 4.65 \times 10^{23}$ J. The volume of the seawater absorbing this energy is $(100 \text{ m})(3.41 \times 10^{14} \text{ m}^2) = 3.41 \times 10^{16} \text{ m}^3$. The mass of this water is $m = \rho V = (1.03 \times 10^3 \text{ kg/m}^3)$ $(3.41 \times 10^{16} \text{ m}^3) = 3.51 \times 10^{19}$ kg. $Q = mc\,\Delta T$, so

$$\Delta T = \frac{Q}{mc} = \frac{4.65 \times 10^{23} \text{ J}}{(3.51 \times 10^{19} \text{ kg})(3890 \text{ J/kg} \cdot \text{C}^\circ)} = 3.4 \text{ C}^\circ$$

Reflect: A temperature rise of 3.4 C$^\circ$ is significant. The solar energy input is a very large number, but so is the total mass of the top 100 m of seawater in the oceans.

15

THERMAL PROPERTIES OF MATTER

Problems 3, 5, 9, 11, 17, 19, 23, 27, 31, 33, 35, 37, 39, 41, 45, 47, 49, 53, 55, 59, 63, 67, 69, 73, 79, 81, 83, 85

Solutions to Problems

15.3. Set Up: $pV = nRT$. $T_1 = 20.0°C = 293$ K.

Solve: (a) n, R, and V are constant. $\dfrac{p}{T} = \dfrac{nR}{V} = $ constant. $\dfrac{p_1}{T_1} = \dfrac{p_2}{T_2}$.

$$T_2 = T_1\left(\frac{p_2}{p_1}\right) = (293 \text{ K})\left(\frac{1.00 \text{ atm}}{3.00 \text{ atm}}\right) = 97.7 \text{ K} = -175°C.$$

(b) $p_2 = 1.00$ atm, $V_2 = 3.00$ L. $p_3 = 3.00$ atm. n, R, and T are constant so $pV = nRT = $ constant. $p_2V_2 = p_3V_3$.

$$V_3 = V_2\left(\frac{p_2}{p_3}\right) = (3.00 \text{ L})\left(\frac{1.00 \text{ atm}}{3.00 \text{ atm}}\right) = 1.00 \text{ L}.$$

Reflect: The final volume is one-third the initial volume. The initial and final pressures are the same, but the final temperature is one-third the initial temperature.

15.5. Set Up: $M = 32.0 \times 10^{-3}$ kg/mol. $T = 22.0°C = 295$ K. $p = 1.00$ atm $= 1.013 \times 10^5$ Pa.

Solve: (a) $n = \dfrac{pV}{RT} = \dfrac{(1.013 \times 10^5 \text{ Pa})(7.00 \text{ m})(8.00 \text{ m})(2.50 \text{ m})}{(8.315 \text{ J/mol} \cdot \text{K})(295 \text{ K})} = 5.78 \times 10^3$ mol

(b) $m_{tot} = nM = (5.78 \times 10^3 \text{ mol})(32.0 \times 10^{-3} \text{ kg/mol}) = 185$ kg.

Reflect: In the ideal gas law it is mandatory that the temperature be in kelvins. Te weight of this mass of gas is about 410 lb.

15.9. Set Up: $pV = nRT$. $T_1 = 300$ K. For thermal expansion of a metal, $\beta = 1 \times 10^{-4} \text{ (C}°)^{-1}$ or less. $R = 0.08206$ L $\cdot$ atm/mol $\cdot$ K.

Solve: (a) The initial pressure is

$$p_1 = \frac{nRT_1}{V_1} = \frac{(11.0 \text{ mol})(0.08206 \text{ L} \cdot \text{atm/mol} \cdot \text{K})(300 \text{ K})}{3.10 \text{ L}} = 87.4 \text{ atm}.$$

$p_2 = 100$ atm. n, R, V are constant so $\dfrac{p}{T} = \dfrac{nR}{V} = $ constant. $\dfrac{p_1}{T_1} = \dfrac{p_2}{T_2}$ and

$$T_2 = T_1\left(\frac{p_2}{p_1}\right) = (300 \text{ K})\left(\frac{100 \text{ atm}}{87.4 \text{ atm}}\right) = 343 \text{ K} = 70°C$$

(b) $\dfrac{\Delta V}{V_0} = \beta\Delta T \approx (1 \times 10^{-4} \text{ (C}°)^{-1})(43 \text{ C}°) = 4 \times 10^{-3}$. The fractional change in volume is very small and can be neglected.

15-1

Reflect: The fractional change in the volume of the tank due to the increased pressure of the gas is

$$\left|\frac{\Delta V}{V_0}\right| = \frac{\Delta p}{B}.$$

Δp is about 1×10^7 Pa and typical values of β for metals are about 1×10^{11} Pa, so this change in volume is also very small and can be neglected.

15.11. Set Up: $pV = nRT$. $T_1 = 300$ K. $p_1 = 1.01 \times 10^5$ Pa.

$$p_2 = 1.01 \times 10^5 \text{ Pa} + 2.72 \times 10^6 \text{ Pa} = 2.82 \times 10^6 \text{ Pa}.$$

Solve: n, R are constant so $\dfrac{pV}{T} = nR = $ constant. $\dfrac{p_1 V_1}{T_1} = \dfrac{p_2 V_2}{T_2}$ and

$$T_2 = T_1\left(\frac{p_2}{p_1}\right)\left(\frac{V_2}{V_1}\right) = (300 \text{ K})\left(\frac{2.82 \times 10^6 \text{ Pa}}{1.01 \times 10^5 \text{ Pa}}\right)\left(\frac{46.2 \text{ cm}^3}{499 \text{ cm}^3}\right) = 775.5 \text{ K} = 502°\text{C}.$$

Reflect: Even though the pressures enter in a ratio, we must use absolute pressures. The ratio of the absolute pressures is different from the ratio of gauge pressures.

15.17. Set Up: The triple-point temperature for water is 273.16 K = 0.01°C.
Solve: The temperature is less than the triple-point temperature so the solid and vapor phases are in equilibrium. The box contains ice and water vapor but no liquid water.
Reflect: Figure 15.7 in Chapter 15 shows that the fusion curve terminates at the triple point.

15.19. Set Up: The density of water is 1.00×10^3 kg/m^3. 1.00 L $= 1.00 \times 10^{-3}$ m^3. $N_A = 6.022 \times 10^{23}$ molecules/mol. For water, $M = 18 \times 10^{-3}$ kg/mol.
Solve: $m = \rho V = (1.00 \times 10^3 \text{ kg/m}^3)(1.00 \times 10^{-3} \text{ m}^3) = 1.00$ kg.

$$n = \frac{m}{M} = \frac{1.00 \text{ kg}}{18 \times 10^{-3} \text{ kg/mol}} = 55.6 \text{ mol}.$$

$$N = nN_A = (55.6 \text{ mol})(6.022 \times 10^{23} \text{ molecules/mol}) = 3.35 \times 10^{25} \text{ molecules}.$$

Reflect: Note that we converted M to kg/mol.

15.23. Set Up: $M = 32.0 \times 10^{-3}$ kg/mol. $v_{\text{rms}} = \sqrt{\dfrac{3kT}{m}}$.

Solve: (a) $v_{\text{rms}} = \sqrt{\dfrac{3(8.315 \text{ J/mol} \cdot \text{K})(300 \text{ K})}{32.0 \times 10^{-3} \text{ kg}}} = 484$ m/s

(b) $K_{\text{av}} = \frac{3}{2}kT = \left(\frac{3}{2}\right)(1.381 \times 10^{-23} \text{ J/molecule} \cdot \text{K})(300 \text{ K}) = 6.21 \times 10^{-21}$ J.
Reflect: K_{av} depends only on T and is independent of the mass of the molecules.

15.27. Set Up: $v_{\text{rms}} = \sqrt{3RT/M}$. $T = 20°\text{C} = 293$ K. $M_{\text{H}_2} = 2.02 \times 10^{-3}$ kg/mol. $M_{\text{O}_2} = 32.0 \times 10^{-3}$ kg/mol. $M_{\text{N}_2} = 28.0 \times 10^{-3}$ kg/mol

Solve: (a) $v_{\text{rms}} = \sqrt{\dfrac{3(8.315 \text{ J/mol} \cdot \text{K})(293 \text{ K})}{2.02 \times 10^{-3} \text{ kg/mol}}} = 1.90 \times 10^3 \text{ m/s} = 1.90$ km/s

(b) $v_{\text{rms}} < 11$ km/s so no, the average H_2 molecule is not moving fast enough to escape.

(c) O_2: $v_{\text{rms,O}_2} = v_{\text{rms,H}_2}\sqrt{\dfrac{M_{\text{H}_2}}{M_{\text{O}_2}}} = (1.90 \text{ km/s})\sqrt{\dfrac{2.02 \times 10^{-3} \text{ kg/mol}}{32.0 \times 10^{-3} \text{ kg/mol}}} = 0.477$ km/s

N_2: $v_{\text{rms,N}_2} = v_{\text{rms,H}_2}\sqrt{\dfrac{M_{\text{H}_2}}{M_{\text{N}_2}}} = (1.90 \text{ km/s})\sqrt{\dfrac{2.02 \times 10^{-3} \text{ kg/mol}}{28.0 \times 10^{-3} \text{ kg/mol}}} = 0.510$ km/s

v_{rms} for H_2 is about four times larger.

(d) Figure 15.12 in Chapter 15 illustrates that some molecules move faster than v_{rms}, so some have a speed greater than the escape speed and can escape. Since the rms speed for H_2 is greater than that of O_2 and N_2, a higher percent of H_2 molecules are moving fast enough to escape.

Reflect: At a given temperature all three species of molecules have the same average kinetic energy. To achieve this, the lighter H_2 molecules have a larger v_{rms}.

15.31. Set Up: $20.9\% - 16.3\% = 4.6\%$ of the volume of air breathed in is O_2 that is absorbed by the body. $T = 20°C = 293$ K. $p = 1$ atm $= 1.013 \times 10^5$ Pa. 1.0 L $= 1.0 \times 10^{-3}$ m^3.

Solve: (a) In one minute the person must absorb

$$\frac{14.5 \text{ L/hr}}{60 \text{ min/hr}} = 0.242 \text{ L/min of } O_2.$$

Since the volume of O_2 absorbed is 4.6% of the air breathed in, the person must have breathed in

$$\frac{0.242 \text{ L/min}}{0.046} = 5.26 \text{ L/m of air.}$$

The number of breaths per minute is

$$\frac{5.26 \text{ L/min}}{0.50 \text{ L/breath}} = 10.5 \text{ breaths/min.}$$

(b) With each breath the person inhales $(0.209)(0.50 \text{ L}) = 0.1045$ L of O_2. $pV = NkT$ gives the number of O_2 molecules in this volume to be

$$N = \frac{pV}{kT} = \frac{(1.013 \times 10^5 \text{ Pa})(0.1045 \times 10^{-3} \text{ m}^3)}{(1.381 \times 10^{-23} \text{ J/molecule} \cdot \text{K})(293 \text{ K})} = 2.62 \times 10^{21} \text{ molecules}$$

Reflect: $V = \dfrac{NkT}{p}$ says that the volume of O_2 is directly proportional to the number of O_2 molecules. About 10 breaths per minute is what we observe for a healthy, resting person; this result makes sense.

15.33. Set Up: The rigid container means the process occurs at constant volume. For a monatomic ideal gas $C_V = \frac{3}{2}R$ and for a diatomic ideal gas $C_V = \frac{5}{2}R$.

Solve: (a) $Q = nC_V \Delta T$ so

$$\Delta T = \frac{Q}{nC_V} = \frac{1850 \text{ J}}{(2.25 \text{ mol})(\frac{5}{2})(8.315 \text{ J/mol} \cdot \text{K})} = 39.6 \text{ C}°.$$

$T_f = T_i + \Delta T = 49.6°C$

(b) $Q = nC_V \Delta T = (2.25 \text{ mol})(\frac{3}{2})(8.315 \text{ J/mol} \cdot \text{K})(39.6 \text{ C}°) = 1110 \text{ J}$

(c) V is constant. Then $pV = nRT$ says p increases when T increases. The process for either gas is sketched in Figure 15.33.

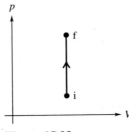

Figure 15.33

Reflect: More heat is required for the diatomic gas because not all of the energy added by Q goes into the translational kinetic energy that determines the temperature. Some of the energy that flows into the gas goes into the rotational kinetic energy of the diatomic molecule. The answer in (b) is smaller than 1850 J by the ratio

$$\frac{\frac{3}{2}R}{\frac{5}{2}R} = \frac{3}{5}.$$

15.35. Set Up: The rule of Dulong and Petit says that the molar heat capacity is $C_V = 3R = 24.9 \text{ J/mol} \cdot \text{K}$. The molar mass of copper is $63.55 \times 10^{-3} \text{ kg/mol}$.

Solve: (a) By comparing $Q = nC \, \Delta T$ and $Q = mc \, \Delta T$ we have $nC = mc$. Then $m = nM$ gives $C = Mc$. So, the rule of Dulong and Petit says that for copper

$$c = \frac{C}{M} = \frac{24.9 \text{ J/mol} \cdot \text{K}}{63.55 \times 10^{-3} \text{ kg/mol}} = 392 \text{ J/kg} \cdot \text{K}$$

(b) The value in Table 14.3 is $c = 390 \text{ J/kg} \cdot \text{K}$. The two values agree well because copper is an elemental solid, composed of individual atoms.

Reflect: Even though the molar heat capacities are the same for all elemental solids, the specific heat capacities are different. For different elements the mass of one mole is different.

15.37. Set Up: The volume is constant.
Solve: (a) The pV diagram is given in Figure 15.37.

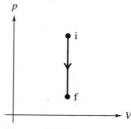

Figure 15.37

(b) Since $\Delta V = 0$, $W = 0$.
Reflect: For any constant volume process the work done is zero.

15.39. Set Up: For a constant pressure process, $W = p \, \Delta V$. If the process is also for an ideal gas, $pV = nRT$ and $p \, \Delta V = nR \, \Delta T$.
Solve: (a) The pV diagram is sketched in Figure 15.39. When the gas is heated at constant pressure, the volume increases.

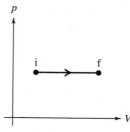

Figure 15.39

(b) $W = nR \, \Delta T = (2.00 \text{ mol})(8.315 \text{ J/mol} \cdot \text{K})(80 \text{ K}) = 1330 \text{ J}.$
Reflect: Since the volume increases, the work done by the gas is positive.

15.41. Set Up: For a constant volume process, $W = 0$. For a constant pressure process, $W = p\,\Delta V$. 1 atm $= 1.013 \times 10^5$ Pa and 1 L $= 10^{-3}$ m^3. The work done by the system in a process is positive when the volume increases and negative when the volume decreases.

Solve: (a) Find W for each process in the cycle.

$1 \to 2$: $W = p\,\Delta V = (2.5 \text{ atm})(1.013 \times 10^5 \text{ Pa/atm})(8 \text{ L} - 2 \text{ L})(10^{-3} \text{ m}^3/1 \text{ L}) = 1.5 \times 10^3$ J

$2 \to 3$: $W = 0$ since $\Delta V = 0$

$3 \to 4$: $W = p\,\Delta V = (0.5 \text{ atm})(1.013 \times 10^5 \text{ Pa/atm})(2 \text{ L} - 8 \text{ L})(10^{-3} \text{ m}^3/1 \text{ L}) = -3.0 \times 10^2$ J

$4 \to 1$: $W = 0$ since $\Delta V = 0$

$$W_{\text{cycle}} = 1.5 \times 10^3 \text{ J} + 0 + (-3.0 \times 10^2 \text{ J}) + 0 = 1.2 \times 10^3 \text{ J}$$

The area enclosed by the cycle is

$$(2.0 \text{ atm})(6.0 \text{ L}) = (12.0 \text{ L} \cdot \text{atm})(10^{-3} \text{ m}^3/\text{L})(1.013 \times 10^5 \text{ Pa/atm}) = 1.2 \times 10^3 \text{ J}.$$

The total work done does equal the area enclosed by the cycle.

(b) For $1 \to 4 \to 3 \to 2 \to 1$, $W_{\text{cycle}} = -1.5 \times 10^3 \text{ J} + 3.0 \times 10^2 \text{ J} = -1.2 \times 10^3$ J. The negative work done for $2 \to 1$ is greater in magnitude than the positive work done in $4 \to 3$ and the total work for the cycle has the opposite sign from what it had in (a).

(c) In the pV-diagram the cycle $3 \to 4 \to 2 \to 3$ is a triangle with area $\frac{1}{2}(2.0 \text{ atm})(6.0 \text{ L}) = 6.0 \text{ L} \cdot \text{atm} = 6.0 \times 10^2$ J. The work done is $\frac{1}{2}$ that done in the cycle of part (a). The positive work done in process $4 \to 2$ is larger in magnitude than the negative work done in process $3 \to 4$ so the net work done in the cycle is positive.

Reflect: For the cycles in parts (a) and (b), the initial and final states are the same but the work done is different. This illustrates the general result that the work done in a process depends not only on the initial and final states but also on the path.

15.45. Set Up: $\Delta U = Q - W$. $Q > 0$ if heat flows into the gas. For a constant pressure process, $W = p\,\Delta V$.

Solve: (a) $W = p\,\Delta V = (2.30 \times 10^5 \text{ Pa})(1.20 \text{ m}^3 - 1.70 \text{ m}^3) = -1.15 \times 10^5$ J

(b) $Q = \Delta U + W = -1.40 \times 10^5 \text{ J} + (-1.15 \times 10^5 \text{ J}) = -2.55 \times 10^5$ J. $Q < 0$ so this amount of heat flows out of the gas.

Reflect: $\Delta V < 0$ and $W < 0$. The work done on the gas adds energy to it, but more energy flows out as heat than is added by the work and the internal energy decreases.

15.47. Set Up: $\Delta U = Q - W$. ΔU is path independent. $Q > 0$ when the system absorbs heat.

Solve: (a) Use path acb to find $\Delta U = U_b - U_a$. $U_b - U_a = Q - W = 90.0 \text{ J} - 60.0 \text{ J} = 30.0$ J. For path adb, $\Delta U = 30.0$ J and $Q = \Delta U + W = 30.0 \text{ J} + 15.0 \text{ J} = 45.0$ J.

(b) $\Delta U = U_a - U_b = -30.0$ J. For process $b \to a$, $\Delta V < 0$ and $W = -35.0$ J. $Q = \Delta U + W = -30.0 \text{ J} + (-35.0 \text{ J}) = -65.0$ J. The system liberated 65.0 J of heat.

(c) For process $a \to d$: $\Delta U = U_d - U_a = 8.0$ J. $W_{d \to b} = 0$ so $W_{a \to d} = W_{adb} = 15.0$ J. $Q = \Delta U + W = 8.0 \text{ J} + 15.0 \text{ J} = 23.0$ J.

For process $d \to b$: $\Delta U = U_b - U_d = 30.0 \text{ J} - 8.0 \text{ J} = 22.0$ J and $W = 0$. $Q = \Delta U + W = 22.0$ J.

Reflect: Both Q and W depend on the path.

15.49. Set Up: For an ideal gas, $\Delta U = nC_V\,\Delta T$ so for an isothermal process $\Delta U = 0$. Use $W = nRT\ln(V_2/V_1)$ to calculate W for the isothermal process and then apply $\Delta U = Q - W$.

Solve: (a) $W = nRT\ln(V_2/V_1) = (1.75 \text{ mol})(8.315 \text{ J/mol} \cdot \text{K})(273 \text{ K})\ln(1.35 \text{ L}/4.20 \text{ L}) = -4510$ J

(b) $\Delta U = Q - W$ and $\Delta U = 0$ says $Q = W = -4510$ J. 4510 J of heat comes out of the gas.

Reflect: When the gas is compressed, heat must be removed to keep the temperature constant.

15.53. Set Up: The pV diagram shows that in the process the volume decreases while the pressure is constant. $1 \text{ L} = 10^{-3} \text{ m}^3$ and $1 \text{ atm} = 1.013 \times 10^5 \text{ Pa}$

Solve: (a) $pV = nRT$. n, R, and p are constant so $\dfrac{V}{T} = \dfrac{nR}{p} = \text{constant.}$ $\dfrac{V_a}{T_a} = \dfrac{V_b}{T_b}$.

$$V_b = V_a \left(\frac{T_b}{T_a}\right) = (0.500 \text{ L})\left(\frac{T_a/4}{T_a}\right) = 0.125 \text{ L}$$

(b) For a constant pressure process, $W = p \, \Delta V = (1.50 \text{ atm})(0.125 \text{ L} - 0.500 \text{ L})$ and

$$W = (-0.5625 \text{ L} \cdot \text{atm})\left(\frac{10^{-3} \text{ m}^3}{1 \text{ L}}\right)\left(\frac{1.013 \times 10^5 \text{ Pa}}{1 \text{ atm}}\right) = -57.0 \text{ J}$$

W is negative since the volume decreases. Since W is negative, work is done on the gas.
(c) For an ideal gas, $U = nCT$ so U decreases when T decreases. The internal energy of the gas decreases because the temperature decreases.
(d) For a constant pressure process, $Q = nC_p \, \Delta T$. T decreases so ΔT is negative and Q is therefore negative. Negative Q means heat leaves the gas.
Reflect: $W = nR \, \Delta T$ and $Q = nC_p \, \Delta T$. $C_p > R$, so more energy leaves as heat than is added by work done on the gas, and the internal energy of the gas decreases.

15.55. Set Up: $pV = nRT$ determines the Kelvin temperature of the gas. The work done in the process is the area under the curve in the pV diagram. Q is positive since heat goes into the gas. $1 \text{ atm} = 1.013 \times 10^5 \text{ Pa}$. $1 \text{ L} = 1 \times 10^{-3} \text{ m}^3$. $\Delta U = Q - W$
Solve: (a) The lowest T occurs when pV has its smallest value. This is at point a, and

$$T_a = \frac{p_a V_a}{nR} = \frac{(0.20 \text{ atm})(1.013 \times 10^5 \text{ Pa/atm})(2.0 \text{ L})(1.0 \times 10^{-3} \text{ m}^3/\text{L})}{(0.0175 \text{ mol})(8.315 \text{ J/mol} \cdot \text{K})} = 278 \text{ K}$$

(b) a to b: $\Delta V = 0$ so $W = 0$
b to c: The work done by the gas is positive since the volume increases. The magnitude of the work is the area under the curve so $W = \frac{1}{2}(0.50 \text{ atm} + 0.30 \text{ atm})(6.0 \text{ L} - 2.0 \text{ L})$ and

$$W = (1.6 \text{ L} \cdot \text{atm})(1 \times 10^{-3} \text{ m}^3/\text{L})(1.013 \times 10^5 \text{ Pa/atm}) = 162 \text{ J}$$

(c) For abc, $W = 162 \text{ J}$. $\Delta U = Q - W = 215 \text{ J} - 162 \text{ J} = 53 \text{ J}$
Reflect: 215 J of heat energy went into the gas. 53 J of energy stayed in the gas as increased internal energy and 162 J left the gas as work done by the gas on its surroundings.

15.59. Set Up: O_2 is diatomic and if treated as ideal it has $C_V = \frac{5}{2}R$ and $C_p = \frac{7}{2}R$. $pV = nRT$. $1 \text{ atm} = 1.013 \times 10^5 \text{ Pa}$. $Q = nC_V \, \Delta T$ for $\Delta V = 0$ and $Q = nC_p \, \Delta T$ for $\Delta p = 0$.
Solve: (a) point a: $T = \dfrac{pV}{RT} = \dfrac{(0.60 \text{ atm})(1.013 \times 10^5 \text{ Pa/atm})(0.10 \text{ m}^3)}{(1.10 \text{ mol})(8.315 \text{ J/mol} \cdot \text{K})} = 665 \text{ K}$

point b: $T = \dfrac{pV}{RT} = 4(665 \text{ K}) = 2660 \text{ K}$

point c: $T = \dfrac{pV}{RT} = \dfrac{2660 \text{ K}}{3} = 887 \text{ K};$

point d: $T = \dfrac{pV}{RT} = \dfrac{887 \text{ K}}{4} = 222 \text{ K}$

(b) and (c) (i) *ab*

$$Q = nC_p \, \Delta T = n(\tfrac{7}{2}R) \, \Delta T = (1.10 \text{ mol})(\tfrac{7}{2})(8.315 \text{ J/mol} \cdot \text{K})(2660 \text{ K} - 665 \text{ K}) = 6.39 \times 10^4 \text{ J}$$

$Q > 0$ so heat enters the gas.

(ii) *bc*

$$Q = nC_V \, \Delta T = n(\tfrac{5}{2}R) \, \Delta T = (1.10 \text{ mol})(\tfrac{5}{2})(8.315 \text{ J/mol} \cdot \text{K})(887 \text{ K} - 2660 \text{ K}) = -4.05 \times 10^4 \text{ J}$$

$Q < 0$ so heat leaves the gas.

(iii) *cd*

$$Q = nC_p \, \Delta T = n(\tfrac{7}{2}R) \, \Delta T = (1.10 \text{ mol})(\tfrac{7}{2})(8.315 \text{ J/mol} \cdot \text{K})(222 \text{ K} - 887 \text{ K}) = -2.13 \times 10^4 \text{ J}$$

$Q < 0$ so heat leaves the gas.

(iv) *da*

$$Q = nC_V \, \Delta T = n(\tfrac{5}{2}R) \, \Delta T = (1.10 \text{ mol})(\tfrac{5}{2})(8.315 \text{ J/mol} \cdot \text{K})(665 \text{ K} - 222 \text{ K}) = 1.01 \times 10^4 \text{ J}$$

$Q > 0$ so heat enters the gas.

Reflect: The net heat flow for the complete cycle is $Q_{ab} + Q_{bc} + Q_{cd} + Q_{da} = 1.22 \times 10^4$ J. The work done in the cycle is positive and equal to the area enclosed by the cycle, so $W = (0.40 \text{ atm})(1.013 \times 10^5 \text{ Pa/atm})(0.30 \text{ m}^3) = 1.22 \times 10^4$ J. For a cycle, where the system returns to the initial state, $\Delta U = 0$ so $Q = W$.

15.63. Set Up: For a monatomic ideal gas, $C_V = \tfrac{3}{2}R$. For an adiabatic process, $Q = 0$, so $\Delta U = -W$. For any process of an ideal gas, $\Delta U = nC_V \, \Delta T$.

Solve: (a) The pV diagram is sketched in Figure 15.63. In an adiabatic expansion of an ideal gas, the pressure drops.

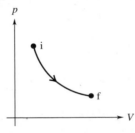

Figure 15.63

(b) $W = -nC_V \, \Delta T = -(0.450 \text{ mol})(\tfrac{3}{2})(8.315 \text{ J/mol} \cdot \text{K})(-40.0 \text{ K}) = +224$ J

(c) $\Delta U = -W = -224$ J.

Reflect: The gas does positive work and this transfers energy out of the gas. $Q = 0$, so the internal energy of the gas decreases.

15.67. Set Up: The pressure difference between two points in a fluid is $\Delta p = \rho g h$, where h is the difference in height of two points.

Solve: (a) $\Delta p = \rho g h = (1.2 \text{ kg/m}^3)(9.80 \text{ m/s}^2)(1000 \text{ m}) = 1.18 \times 10^4$ Pa

(b) At the bottom of the mountain, $p = 1.013 \times 10^5$ Pa. At the top, $p = 8.95 \times 10^4$ Pa. $pV = nRT = $ constant so $p_b V_b = p_t V_t$ and

$$V_t = V_b \left(\frac{p_b}{p_t} \right) = (0.50 \text{ L}) \left(\frac{1.013 \times 10^5 \text{ Pa}}{8.95 \times 10^4 \text{ Pa}} \right) = 0.566 \text{ L}.$$

Reflect: The pressure variation with altitude is affected by changes in air density and temperature and we have neglected those effects. The pressure decreases with altitude and the volume increases. You may have noticed this effect: bags of potato chips "puff-up" when taken to the top of a mountain.

15.69. Set Up: In $pV = nRT$ we must use the absolute pressure. $T_1 = 278$ K. $p_1 = 2.72$ atm. $T_2 = 318$ K.

Solve: n, R constant, so $\dfrac{pV}{T} = nR =$ constant. $\dfrac{p_1 V_1}{T_1} = \dfrac{p_2 V_2}{T_2}$ and

$$p_2 = p_1 \left(\frac{V_1}{V_2}\right)\left(\frac{T_2}{T_1}\right) = (2.72 \text{ atm})\left(\frac{0.0150 \text{ m}^3}{0.0159 \text{ m}^3}\right)\left(\frac{318 \text{ K}}{278 \text{ K}}\right) = 2.94 \text{ atm}.$$

The final gauge pressure is 2.94 atm $-$ 1.02 atm $=$ 1.92 atm.

Reflect: Since a ratio is used, pressure can be expressed in atm. But absolute pressures must be used. The ratio of gauge pressures is not equal to the ratio of absolute pressures.

15.73. Set Up: For helium, $C_V = 12.47$ J/mol $\cdot$ K and $C_p = 20.78$ J/mol $\cdot$ K.
Solve: (a) $\Delta T = 0$. $\Delta U = 0$ and $Q = W = 300$ J.
(b) $Q = 0$. $\Delta U = -W = -300$ J.
(c) $\Delta p = 0$. $W = p \, \Delta V = nR \, \Delta T$.

$$Q = nC_p \, \Delta T = \frac{C_p}{R} W = \left(\frac{20.78 \text{ J/mol} \cdot \text{K}}{8.315 \text{ J/mol} \cdot \text{K}}\right)(300 \text{ J}) = 750 \text{ J}.$$

$\Delta U = Q - W = 750$ J $-$ 300 J $=$ 450 J. This is also $\left(\dfrac{C_V}{R}\right)W$.

Reflect: In any process of an ideal gas, $\Delta U = nC_V \, \Delta T$. ΔU is different for each of these processes because they have different values of ΔT.

15.79. Set Up: ab is at constant volume, ca is at constant pressure and bc is at constant temperature. For $\Delta T = 0$, $\Delta U = 0$ and $Q = W = nRT\ln(V_c/V_b)$. For ideal H_2 (diatomic), $C_V = \frac{5}{2}R$ and $C_p = \frac{7}{2}R$. $\Delta U = nC_V \, \Delta T$ for any process of an ideal gas.
Solve: (a) $T_b = T_c$. For states b and c, $pV = nRT =$ constant so $p_b V_b = p_c V_c$ and

$$V_c = V_b\left(\frac{p_b}{p_c}\right) = (0.20 \text{ L})\left(\frac{2.0 \text{ atm}}{0.50 \text{ atm}}\right) = 0.80 \text{ L}$$

(b) $T_a = \dfrac{p_a V_a}{nR} = \dfrac{(0.50 \text{ atm})(1.013 \times 10^5 \text{ Pa/atm})(0.20 \times 10^{-3} \text{ m}^3)}{(0.0040 \text{ mol})(8.315 \text{ J/mol} \cdot \text{K})} = 305 \text{ K}$

$V_a = V_b$ so for states a and b, $\dfrac{T}{p} = \dfrac{V}{nR} =$ constant so $\dfrac{T_a}{p_a} = \dfrac{T_b}{p_b}$.

$$T_b = T_c = T_a\left(\frac{p_b}{p_a}\right) = (305 \text{ K})\left(\frac{2.0 \text{ atm}}{0.50 \text{ atm}}\right) = 1220 \text{ K}; \, T_c = 1220 \text{ K}$$

(c) ab: $Q = nC_V \, \Delta T = n(\frac{5}{2}R) \, \Delta T = (0.0040 \text{ mol})(\frac{5}{2})(8.315 \text{ J/mol} \cdot \text{K})(1220 \text{ K} - 305 \text{ K}) = +76$ J
Q is positive and heat goes into the gas.
ca: $Q = nC_p \, \Delta T = n(\frac{7}{2}R) \, \Delta T = (0.0040 \text{ mol})(\frac{7}{2})(8.315 \text{ J/mol} \cdot \text{K})(305 \text{ K} - 1220 \text{ K}) = -107$ J
Q is negative and heat comes out of the gas.
bc: $Q = W = nRT\ln(V_c/V_b) = (0.0040 \text{ mol})(8.315 \text{ J/mol} \cdot \text{K})(1220 \text{ K})\ln(0.80 \text{ L}/0.20 \text{ L}) = 56$ J
Q is positive and heat goes into the gas.
(d) ab: $\Delta U = nC_V \, \Delta T = n(\frac{5}{2}R) \, \Delta T = (0.0040 \text{ mol})(\frac{5}{2})(8.315 \text{ J/mol} \cdot \text{K})(1220 \text{ K} - 305 \text{ K}) = +76$ J
The internal energy increased.
bc: $\Delta T = 0$ so $\Delta U = 0$. The internal energy does not change.
ca: $\Delta U = nC_V \, \Delta T = n(\frac{5}{2}R) \, \Delta T = (0.0040 \text{ mol})(\frac{5}{2})(8.315 \text{ J/mol} \cdot \text{K})(305 \text{ K} - 1220 \text{ K}) = -76$ J
The internal energy decreased.
Reflect: The net internal energy change for the complete cycle $a \to b \to c \to a$ is $\Delta U_{\text{tot}} = +76$ J $+ 0 +$ $(-76$ J$) = 0$. For any complete cycle the final state is the same as the initial state and the net internal energy change is zero. For the cycle the net heat flow is $Q_{\text{tot}} = +76$ J $+ (-107$ J$) + 56$ J $= +25$ J. $\Delta U_{\text{tot}} = 0$ so $Q_{\text{tot}} = W_{\text{tot}}$. The net work done in the cycle is positive and this agrees with our result that the net heat flow is positive.

15.81. Set Up: $pV = nRT$ and $m = nM$. We must use absolute pressure in $pV = nRT$. $p_1 = 4.01 \times 10^5$ Pa, $p_2 = 2.81 \times 10^5$ Pa. $T_1 = 310$ K, $T_2 = 295$ K.

Solve: (a) $n_1 = \dfrac{p_1 V_1}{RT_1} = \dfrac{(4.01 \times 10^5 \text{ Pa})(0.075 \text{ m}^3)}{(8.315 \text{ J/mol} \cdot \text{K})(310 \text{ K})} = 11.7$ mol.

$$m = nM = (11.7 \text{ mol})(32.0 \text{ g/mol}) = 374 \text{ g.}$$

(b) $n_2 = \dfrac{p_2 V_2}{RT_2} = \dfrac{(2.81 \times 10^5 \text{ Pa})(0.075 \text{ m}^3)}{(8.315 \text{ J/mol} \cdot \text{K})(295 \text{ K})} = 8.59$ mol.

$m = 275$ g. The mass that has leaked out is $374 \text{ g} - 275 \text{ g} = 99$ g.

Reflect: In the ideal gas law we must use absolute pressure and T must be in kelvins

15.83. Set Up: The initial absolute pressure is 2.39×10^5 Pa and the final absolute pressure is 1.24×10^5 Pa. For a constant pressure process, $W = p\,\Delta V$. For an adiabatic process, $Q = 0$ and $W = -\Delta U = -nC_V \Delta T$. The two processes are shown in Figure 15.83. The gas does positive work when it expands.

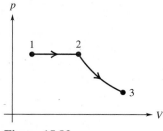

Figure 15.83

Solve: Constant pressure process $1 \rightarrow 2$.

$$W = p\,\Delta V = (2.39 \times 10^5 \text{ Pa})(1.42 \text{ m}^3 - 0.28 \text{ m}^3) = 2.72 \times 10^5 \text{ J}$$

Adiabatic process $2 \rightarrow 3$. $pV = nRT$ so $T_2 = \dfrac{p_2 V_2}{nR}$ and $T_3 = \dfrac{p_3 V_3}{nR}$. $W = -\dfrac{C_V}{R}(p_3 V_3 - p_2 V_2)$

$$W = -\left(\frac{20.8 \text{ J/mol} \cdot \text{K}}{8.315 \text{ J/mol} \cdot \text{K}}\right)([1.24 \times 10^5 \text{ Pa}][2.27 \text{ m}^3] - [2.39 \times 10^5 \text{ Pa}][1.42 \text{ m}^3]) = 1.45 \times 10^5 \text{ J.}$$

The total work done by the gas is $2.72 \times 10^5 \text{ J} + 1.45 \times 10^5 \text{ J} = 4.17 \times 10^5 \text{ J.}$

Reflect: Each process is an expansion, so the work done during each is positive.

15.85. Set Up: $T_1 = 300$ K. When the volume doubles at constant pressure the temperature doubles, so $T_2 = 600$ K. For helium, $C_p = 20.78$ J/mol $\cdot$ K and $\gamma = 1.67$. $\Delta U = nC_V \Delta T$ for any process of an ideal gas. $\Delta U = Q - W$.

Solve: (a) The process is sketched in Figure 15.85.

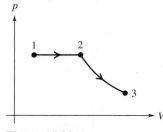

Figure 15.85

(b) For the isobaric step, $Q = nC_p \Delta T = (2.00 \text{ mol})(20.78 \text{ J/mol} \cdot \text{K})(300 \text{ K}) = 1.25 \times 10^4$ J. For the adiabatic process, $Q = 0$. The total Q is 1.25×10^4 J.

(c) $\Delta U = 0$ since $\Delta T = 0$.

(d) Since $\Delta U = 0$, $W = Q = 1.25 \times 10^4$ J.

(e) $T_3 = 300$ K, $T_2 = 600$ K and $V_2 = 0.0600$ m^3. $T_2 V_2^{\gamma-1} = T_3 V_3^{\gamma-1}$.

$$V_3 = V_2 \left(\frac{T_2}{T_3}\right)^{1/(\gamma-1)} = (0.0600 \text{ m}^3)\left(\frac{600 \text{ K}}{300 \text{ K}}\right)^{1/0.67} = 0.169 \text{ m}^3.$$

Reflect: In both processes the internal energy changes. In the isobaric expansion the temperature increases and the internal energy increases. In the adiabatic expansion the temperature decreases and $\Delta U < 0$. The magnitude of the two temperature changes are equal and the net change in internal energy is zero.

16

THE SECOND LAW OF THERMODYNAMICS

Problems 3, 7, 9, 13, 15, 17, 21, 25, 27, 29, 31, 33, 39, 43, 45, 47, 49, 55, 57, 59

Solutions to Problems

16.3. Set Up: For a heat engine, $W = |Q_H| - |Q_C|$. $e = \dfrac{W}{Q_H}$. $Q_H > 0$, $Q_C < 0$.

Solve: (a) $W = 2200$ J. $|Q_C| = 4300$ J. $Q_H = W + |Q_C| = 6500$ J.

(b) $e = \dfrac{2200 \text{ J}}{6500 \text{ J}} = 0.34 = 34\%$.

Reflect: Since the engine operates on a cycle, the net Q equal the net W. But to calculate the efficiency we use the heat energy input, Q_H

16.7. Set Up: $W = |Q_H| - |Q_C|$. $e = \dfrac{W}{Q_H}$. $Q_H > 0$, $Q_C < 0$.

Solve: (a) $e = \dfrac{W}{Q_H} = \dfrac{W/t}{Q_H/t} = \dfrac{330 \text{ MW}}{1300 \text{ MW}} = 0.25 = 25\%$.

(b) $|Q_C| = |Q_H| - W$ so $|Q_C|/t = |Q_H|/t - W/t = 1300 \text{ MW} - 330 \text{ MW} = 970 \text{ MW}$.

Reflect: The equations for e and W have the same form when written in terms of power output and rate of heat flow.

16.9. Set Up: ca is at constant volume, ab has $Q = 0$, and bc is at constant pressure. For a constant pressure process $W = p \, \Delta V$ and $Q = nC_p \, \Delta T$. $pV = nRT$ gives $n \, \Delta T = \dfrac{p \, \Delta V}{R}$ so $Q = \left(\dfrac{C_p}{R}\right) p \, \Delta V$. If $\gamma = 1.40$ the gas is diatomic and $C_p = \frac{7}{2}R$. For a constant volume process $W = 0$ and $Q = nC_V \, \Delta T$. $pV = nRT$ gives $n \, \Delta T = \dfrac{V \, \Delta p}{R}$ so $Q = \left(\dfrac{C_V}{R}\right) V \, \Delta p$. For a diatomic ideal gas $C_V = \frac{5}{2}R$. 1 atm $= 1.013 \times 10^5$ Pa

Solve: (a) $V_b = 9.0 \times 10^{-3}$ m³, $p_b = 1.5$ atm and $V_a = 2.0 \times 10^{-3}$ m³. For an adiabatic process $p_a V_a^\gamma = p_b V_b^\gamma$.

$$p_a = p_b \left(\dfrac{V_b}{V_a}\right)^\gamma = (1.5 \text{ atm}) \left(\dfrac{9.0 \times 10^{-3} \text{ m}^3}{2.0 \times 10^{-3} \text{ m}^3}\right)^{1.4} = 12.3 \text{ atm}$$

(b) Heat enters the gas in process ca, since T increases.

$$Q = \left(\dfrac{C_V}{R}\right) V \, \Delta p = \left(\tfrac{5}{2}\right) (2.0 \times 10^{-3} \text{ m}^3)(12.3 \text{ atm} - 1.5 \text{ atm})(1.013 \times 10^5 \text{ Pa/atm}) = 5470 \text{ J}$$

$$Q_H = 5470 \text{ J}$$

(c) Heat leaves the gas in process bc, since T increases.

$$Q = \left(\dfrac{C_p}{R}\right) p \, \Delta V = \left(\tfrac{7}{2}\right)(1.5 \text{ atm})(1.013 \times 10^5 \text{ Pa/atm})(-7.0 \times 10^{-3} \text{ m}^3) = -3723 \text{ J}$$

$$Q_C = -3723 \text{ J}$$

(d) $W = Q_H + Q_C = +5470 \text{ J} + (-3723 \text{ J}) = 1747 \text{ J}$

(e) $e = \dfrac{W}{Q_H} = \dfrac{1747 \text{ J}}{5470 \text{ J}} = 0.319 = 31.9\%$

Reflect: We did not use the number of moles of the gas.

16.13. Set Up: Process $a \rightarrow b$ is adiabatic. $T_a = 295 \text{ K}$.

Solve: (a) $T_a V_a^{\gamma-1} = T_b V_b^{\gamma-1}$.

$$T_b = T_a \left(\frac{V_a}{V_b}\right)^{\gamma-1} = (295 \text{ K})\left(\frac{rV}{V}\right)^{0.40} = (295 \text{ K})(9.50)^{0.40} = 726 \text{ K} = 453°\text{C}.$$

(b) $pV = nRT$ gives $\dfrac{p_a V_a}{T_a} = \dfrac{p_b V_b}{T_b}$.

$$p_b = \left(\frac{T_b}{T_a}\right)\left(\frac{V_a}{V_b}\right)p_a = \left(\frac{726 \text{ K}}{295 \text{ K}}\right)(9.50)(8.50 \times 10^4 \text{ Pa}) = 1.99 \times 10^6 \text{ Pa}.$$

Reflect: In an adiabatic compression both the pressure and temperature increase.

16.15. Set Up: For a refrigerator, $|Q_H| = |Q_C| + |W|$. $Q_C > 0$, $Q_H < 0$. $K = \dfrac{|Q_C|}{|W|}$.

Solve: (a) $W = \dfrac{|Q_C|}{K} = \dfrac{3.40 \times 10^4 \text{ J}}{2.10} = 1.62 \times 10^4 \text{ J}$.

(b) $|Q_H| = 3.40 \times 10^4 \text{ J} + 1.62 \times 10^4 \text{ J} = 5.02 \times 10^4 \text{ J}$.

Reflect: More heat is discarded to the high temperature reservoir than is absorbed from the cold reservoir.

16.17. Set Up: $|Q_H| = |Q_C| + |W|$. $K = \dfrac{|Q_C|}{W}$. For water, $c_w = 4190 \text{ J/kg} \cdot \text{K}$ and $L_f = 3.34 \times 10^5 \text{ J/kg}$. For ice, $c_{ice} = 2010 \text{ J/kg} \cdot \text{K}$.

Solve: (a)

$Q = m c_{ice} \Delta T_{ice} - m L_f + m c_w \Delta T_w$

$Q = (1.80 \text{ kg})([2010 \text{ J/kg} \cdot \text{K}][-5.0 \text{ C}°] - 3.34 \times 10^5 \text{ J/kg} + [4190 \text{ J/kg} \cdot \text{K}][-25.0 \text{ C}°]) = -8.08 \times 10^5 \text{ J}$

$Q = -8.08 \times 10^5 \text{ J}$. Q is negative for the water since heat is removed from it.

(b) $|Q_C| = 8.08 \times 10^5 \text{ J}$.

$$W = \frac{|Q_C|}{K} = \frac{8.08 \times 10^5 \text{ J}}{2.40} = 3.37 \times 10^5 \text{ J}.$$

(c) $|Q_H| = 8.08 \times 10^5 \text{ J} + 3.37 \times 10^5 \text{ J} = 1.14 \times 10^6 \text{ J}$.

Reflect: For this device, $Q_C > 0$ and $Q_H < 0$. More heat is rejected to the room than is removed from the water.

16.21. Set Up: $e_{\text{Carnot}} = 1 - \dfrac{T_C}{T_H}$ and $e = \dfrac{W}{Q_H}$. $W = Q_C + Q_H$. For an engine, $Q_H > 0$, $W > 0$ and $Q_C < 0$.

Solve: (a) $T_C = 65°\text{F} = 18.3°\text{C} = 291 \text{ K}$. $e = 1 - \dfrac{T_C}{T_H}$ so

$$T_H = \frac{T_C}{1 - e} = \frac{291 \text{ K}}{1 - 0.67} = 882 \text{ K} = 609°\text{C} = 1128°\text{F}$$

(b) $W = e Q_H = (0.67)(100 \text{ J}) = 67 \text{ J}$ and $Q_C = W - Q_H = 67 \text{ J} - 100 \text{ J} = -33 \text{ J}$.

Reflect: We assumed a Carnot cycle, because that cycle has the greatest possible efficiency for a given set of reservoir temperatures, T_C and T_H. The Carnot cycle is an idealized model with reversible heat flows. Any real engine will have an efficiency less than e_{Carnot} and will require a larger T_H in order to achieve a 67% efficiency.

16.25. Set Up: For a Carnot refrigerator, $K = \dfrac{T_C}{T_H - T_C}$. $T_C = -10 + 273 = 263$ K. $T_H = 20 + 273 = 293$ K.

Also, $K = \dfrac{|Q_C|}{|W|}$, where $|Q_C|$ is the heat that must be removed from the water to cool it to $0°C$ and freeze it. For water, $c = 4.19 \times 10^3$ J/kg $\cdot$ C° and $L_f = 334 \times 10^3$ J/kg. $|Q_H| = |W| + |Q_C|$, where $|Q_H|$ is the heat expelled to the room.

Solve: (a) $K = \dfrac{T_C}{T_H - T_C} = \dfrac{263 \text{ K}}{293 \text{ K} - 263 \text{ K}} = 8.77$

(b) $Q = m c_w \, \Delta T_w - m L_f$

$$Q = (0.375 \text{ kg})([4.19 \times 10^3 \text{ J/kg} \cdot \text{C°}][0°C - 18°C] - 334 \times 10^3 \text{ J/kg}) = -1.54 \times 10^5 \text{ J}$$

1.54×10^5 J must be removed from the water to convert it to ice at $0°C$

(c) $|W| = \dfrac{|Q_C|}{K} = \dfrac{1.54 \times 10^5 \text{ J}}{8.77} = 1.76 \times 10^4 \text{ J}$

(d) $|Q_H| = |W| + |Q_C| = 1.76 \times 10^4 \text{ J} + 1.54 \times 10^5 \text{ J} = 1.72 \times 10^5 \text{ J}$

Reflect: More energy is expelled to the room than is removed from the water. K is larger when T_H is close to T_C; less energy must be expelled to move heat energy through a smaller temperature difference.

16.27. Set Up: For an engine, W and Q_H are positive and Q_C is negative. For a refrigerator, Q_C is positive and W and Q_H are negative. A heat pump air conditioner takes in heat at a cool place and expels heat into a warm place. A heat pump house heater takes in heat at a cool place and expels heat into a warm place. For both types of heat pumps, W is negative and energy must be supplied to operate the device. bc and da are adiabatic, so the heat flow for these processes is zero.

Solve: (a) The cycle is clockwise. More positive work is done during ab and bc than the magnitude of the negative work done in cd and da, so the net work done in the cycle is positive. Heat enters the gas in ab and leaves the gas in cd.

(b) The cycle is counterclockwise and the net work done in the cycle is negative. Heat enters the gas in dc and leaves the gas in ba. dc occurs inside the food compartment and ba occurs in the air of the room.

(c) The cycle is counterclockwise and the net work done in the cycle is negative. Heat enters the gas in dc and leaves the gas in ba. dc occurs inside the house and ba occurs outside.

(d) The cycle is counterclockwise and the net work done in the cycle is negative. Heat enters the gas in dc and leaves the gas in ba. dc occurs outside and ba occurs inside the house.

Reflect: For a refrigerator the signs of W, Q_H and Q_C are all opposite to what they are for an engine.

16.29. Set Up: $\Delta S = \dfrac{Q}{T}$ for each object. For water, $L_f = 3.34 \times 10^5$ J/kg.

Solve: (a) The heat flow into the ice is $Q = m L_f = (0.350 \text{ kg})(3.34 \times 10^5 \text{ J/kg}) = 1.17 \times 10^5$ J. The heat flow occurs at $T = 273$ K, so

$$\Delta S = \frac{Q}{T} = \frac{1.17 \times 10^5 \text{ J}}{273 \text{ K}} = 429 \text{ J/K}.$$

Q is positive and ΔS is positive.

(b) $Q = -1.17 \times 10^5$ J flows out of the heat source, at $T = 298$ K.

$$\Delta S = \frac{Q}{T} = \frac{-1.17 \times 10^5 \text{ J}}{298 \text{ K}} = -393 \text{ J/K}.$$

Q is negative and ΔS is negative.

(c) $\Delta S_{tot} = 429 \text{ J/K} + (-393 \text{ J/K}) = +36 \text{ J/K}.$

Reflect: For the total isolated system, $\Delta S > 0$ and the process is irreversible.

16.31. Set Up: $e_{Carnot} = 1 - \dfrac{T_C}{T_H}$. $T_C = 20.0°C = 293$ K and $T_H = 250°C = 523$ K. For any engine, $e = \dfrac{W}{Q_H}$.
$|Q_H| = |W| + |Q_C|$ with $W > 0$, $Q_H > 0$ and $Q_C < 0$.

Solve: **(a)** $e_{Carnot} = 1 - \dfrac{293\text{ K}}{523\text{ K}} = 0.440$. $e = 0.75e_{Carnot} = 0.330$.

$$Q_H = \frac{W}{e} = \frac{375\text{ J}}{0.330} = 1.14 \times 10^3 \text{ J}.$$

$|Q_C| = |Q_H| - |W| = 1140\text{ J} - 375\text{ J} = 765$ J

(b) For the hot reservoir heat $|Q_H|$ flows out of the reservoir at a temperature T_H:

$$\Delta S = \frac{-|Q_H|}{T_H} = \frac{-1.14 \times 10^3 \text{ J}}{523\text{ K}} = -2.18 \text{ J/K}.$$

For the cold reservoir heat $|Q_C|$ flows into the reservoir at a temperature T_C:

$$\Delta S = \frac{+|Q_C|}{T_C} = \frac{765\text{ J}}{293\text{ K}} = +2.61 \text{ J/K}.$$

The total entropy change of the factory is -2.18 J/K $+ 2.61$ J/K $= +0.43$ J/K.
Reflect: For a Carnot cycle the net entropy change of the reservoirs for a cycle is zero. This engine has irreversible steps and the net entropy change for the reservoirs is positive.

16.33. Set Up: The reversible process that connects the same initial and final states is an isothermal expansion at $T = 293$ K, from $V_1 = 10.0$ L to $V_2 = 35.0$ L. For an isothermal expansion of an ideal gas $\Delta U = 0$ and $Q = W = nRT\ln(V_2/V_1)$.
Solve: **(a)** $Q = (3.20\text{ mol})(8.315\text{ J/mol}\cdot\text{K})(293\text{ K})\ln(35.0\text{ L}/10.0\text{ L}) = 9767$ J

$$\Delta S = \frac{Q}{T} = \frac{9767\text{ J}}{293\text{ K}} = +33.3 \text{ J/K}.$$

(b) The isolated system has $\Delta S > 0$ so the process is irreversible.
Reflect: The reverse process, where all the gas in 35.0 L goes through the hole and into the tank doesn't occur.

16.39. Set Up: 1 food-calorie $= 1000$ cal $= 4186$ J. The heat enters the person's body at $37°C = 310$ K and leaves at a temperature of $30°C = 303$ K. $\Delta S = \dfrac{Q}{T}$

Solve: $|Q| = (0.80)(2.50\text{ g})(9.3\text{ food-calorie/g})\left(\dfrac{4186\text{ J}}{1\text{ food-calorie}}\right) = 7.79 \times 10^4$ J

$$\Delta S = \frac{+7.79 \times 10^4 \text{ J}}{310\text{ K}} + \frac{-7.79 \times 10^4 \text{ J}}{303\text{ K}} = -5.8 \text{ J/K}.$$

Your body's entropy decreases.
Reflect: The entropy of your body can decrease without violating the second law of thermodynamics because you are not an isolated system.

16.43. Set Up: For water, $c = 4190$ J/kg $\cdot$ K and $\rho = 1.00 \times 10^3$ kg/m³.
Solve: The energy collected in 1 h $= 3600$ s is $Q = (0.50)(200\text{ W/m}^2)(30.0\text{ m}^2)(3600\text{ s}) = 1.08 \times 10^7$ J. The heat goes into the water, so $Q = mc\,\Delta T$.

$$m = \frac{Q}{c\,\Delta T} = \frac{1.08 \times 10^7 \text{ J}}{(4190\text{ J/kg}\cdot\text{K})(45.0\text{ C}°)} = 57.3 \text{ kg}.$$

The volume of water is $V = \dfrac{m}{\rho} = \dfrac{57.3\text{ kg}}{1.00 \times 10^3 \text{ kg/m}^3} = 57.3 \times 10^{-3} \text{ m}^3 = 57.3$ L.
Reflect: This is adequate for a typical household.

16.45. Set Up: For an ideal diatomic gas, $C_V = \frac{5}{2}R$, $C_p = \frac{7}{2}R$ and $\gamma = \dfrac{C_p}{C_V} = \dfrac{7}{5}$. $pV = nRT$. $\Delta U = Q - W$.

Solve: (a) $p_1 = 1.00$ atm.

$$V_1 = \frac{nRT_1}{p_1} = \frac{(0.350 \text{ mol})(0.08206 \text{ L} \cdot \text{atm/mol} \cdot \text{K})(300 \text{ K})}{1.00 \text{ atm}} = 8.62 \text{ L}.$$

$V_2 = V_1 = 8.62$ L. $\dfrac{p_1}{T_1} = \dfrac{p_2}{T_2}$ so

$$p_2 = \left(\frac{T_2}{T_1}\right)p_1 = \left(\frac{600 \text{ K}}{300 \text{ K}}\right)(1.00 \text{ atm}) = 2.00 \text{ atm}.$$

$p_3 = 1.00$ atm.

$$V_3 = \frac{nRT_3}{p_3} = \frac{(0.350 \text{ mol})(0.08206 \text{ L} \cdot \text{atm/mol} \cdot \text{K})(492 \text{ K})}{1.00 \text{ atm}} = 14.1 \text{ L}.$$

(b) $1 \to 2$: $W = 0$ since $\Delta V = 0$.

$$Q = \Delta U = nC_V \Delta T = (0.350 \text{ mol})(\tfrac{5}{2})(8.315 \text{ J/mol} \cdot \text{K})(300 \text{ K}) = 2183 \text{ J}.$$

$2 \to 3$: $Q = 0$. $\Delta U = nC_V \Delta T = (0.350 \text{ mol})(\tfrac{5}{2})(8.315 \text{ J/mol} \cdot \text{K})(-108 \text{ K}) = -786 \text{ J}$. $W = -\Delta U = +786 \text{ J}$.

$3 \to 1$: $W = p \Delta V = nR \Delta T = (0.350 \text{ mol})(8.315 \text{ J/mol} \cdot \text{K})(-192 \text{ K}) = -559 \text{ J}$.

$$Q = nC_p \Delta T = (0.350 \text{ mol})(\tfrac{7}{2})(8.315 \text{ J/mol} \cdot \text{K})(-192 \text{ K}) = -1955 \text{ J}.$$

$$\Delta U = Q - W = -1955 \text{ J} - (-559 \text{ J}) = -1396 \text{ J}.$$

(c) $W_{\text{tot}} = W_{1 \to 2} + W_{2 \to 3} + W_{3 \to 1} = 0 + 786 \text{ J} + (-559 \text{ J}) = +227 \text{ J}$

(d) $Q_{\text{tot}} = Q_{1 \to 2} + Q_{2 \to 3} + Q_{3 \to 1} = 2183 \text{ J} + 0 + (-1955 \text{ J}) = 228 \text{ J}$. $Q_{\text{tot}} = W_{\text{tot}}$, apart from a small difference due to rounding.

(e) $e = \dfrac{W}{Q_H} = \dfrac{W}{Q_{1 \to 2}} = \dfrac{227 \text{ J}}{2183 \text{ J}} = 0.104 = 10.4\%.$

$$e_{\text{Carnot}} = 1 - \frac{T_C}{T_H} = 1 - \frac{300 \text{ K}}{600 \text{ K}} = 0.500 = 50.0\%.$$

$e < e_{\text{Carnot}}$.

Reflect: The net ΔU for the cycle is $\Delta U_{\text{tot}} = 2183 \text{ J} + (-786 \text{ J}) + (-1396 \text{ J})$. This is zero, except for a small difference due to rounding. For any cycle, $\Delta U = 0$.

16.47. Set Up: The least possible $|W|$ and the smallest $|Q_H|$ is for a Carnot cycle. $T_C = -5.0°C = 268$ K. $T_H = 20.0°C = 293$ K. For water, $c_w = 4190 \text{ J/kg} \cdot \text{K}$ and $L_f = 3.34 \times 10^5 \text{ J/kg}$. $|Q_H| = |W| + |Q_C|$. $Q_H < 0$, $Q_C > 0$, $W < 0$. $\dfrac{Q_C}{Q_H} = -\dfrac{T_C}{T_H}$.

Solve: The heat that must be removed from the water is

$$Q = mc \Delta T - mL_f = (5.00 \text{ kg})([4190 \text{ J/kg} \cdot \text{K}][-20.0 \text{ C}°] - 3.34 \times 10^5 \text{ J/kg}) = -2.09 \times 10^6 \text{ J}.$$

This amount of heat goes into the working substance of the freezer, so $Q_C = 2.09 \times 10^6$ J. $\dfrac{Q_C}{Q_H} = -\dfrac{T_C}{T_H}$ so

$$Q_H = -Q_C\frac{T_H}{T_C} = -(2.09 \times 10^6 \text{ J})\left(\frac{293 \text{ K}}{268 \text{ K}}\right) = -2.28 \times 10^6 \text{ J}.$$

$|Q_H| = 2.28 \times 10^6$ J is expelled into the room. $|W| = |Q_H| - |Q_C| = 2.28 \times 10^6 \text{ J} - 2.09 \times 10^6 \text{ J} = 1.9 \times 10^5 \text{ J}$ is the least possible amount of electrical energy that must be supplied.

Reflect: The amount of heat expelled into the room is greater than the amount of heat removed from the water.

16.49. Set Up: For O_2, $C_V = 20.85$ J/mol · K and $C_p = 29.17$ J/mol · K.

Solve: (a) $p_1 = 2.00$ atm, $V_1 = 4.00$ L, $T_1 = 300$ K. $p_2 = 2.00$ atm. $\dfrac{V_1}{T_1} = \dfrac{V_2}{T_2}$.

$$V_2 = \left(\frac{T_2}{T_1}\right)V_1 = \left(\frac{450 \text{ K}}{300 \text{ K}}\right)(4.00) = 6.00 \text{ L}.$$

$V_3 = 6.00$ L. $\dfrac{p_2}{T_2} = \dfrac{p_3}{T_3}$.

$$p_3 = \left(\frac{T_3}{T_2}\right)p_2 = \left(\frac{250 \text{ K}}{450 \text{ K}}\right)(2.00 \text{ atm}) = 1.11 \text{ atm}$$

$V_4 = 4.00$ L. $p_3 V_3 = p_4 V_4$.

$$p_4 = p_3\left(\frac{V_3}{V_4}\right) = (1.11 \text{ atm})\left(\frac{6.00 \text{ L}}{4.00 \text{ L}}\right) = 1.67 \text{ atm}.$$

These processes are shown in Figure 16.49.

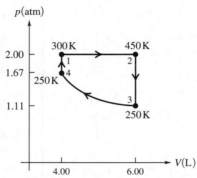

Figure 16.49

(b) $n = \dfrac{p_1 V_1}{RT_1} = \dfrac{(2.00 \text{ atm})(4.00 \text{ L})}{(0.08206 \text{ L} \cdot \text{atm/mol} \cdot \text{K})(300 \text{ K})} = 0.325$ mol

process $1 \rightarrow 2$: $W = p\,\Delta V = nR\,\Delta T = (0.325 \text{ mol})(8.315 \text{ J/mol} \cdot \text{K})(150 \text{ K}) = 405$ J.

$$Q = nC_p\,\Delta T = (0.325 \text{ mol})(29.17 \text{ J/mol} \cdot \text{K})(150 \text{ K}) = 1422 \text{ J}.$$

process $2 \rightarrow 3$: $W = 0$. $Q = nC_V\,\Delta T = (0.325 \text{ mol})(20.85 \text{ J/mol} \cdot \text{K})(-200 \text{ K}) = -1355$ J.

process $3 \rightarrow 4$: $\Delta U = 0$ and

$$Q = W = nRT_3\ln\left(\frac{V_4}{V_3}\right) = (0.325 \text{ mol})(8.315 \text{ J/mol} \cdot \text{K})(250 \text{ K})\ln\left(\frac{4.00 \text{ L}}{6.00 \text{L}}\right) = -274 \text{ J}.$$

process $4 \rightarrow 1$: $W = 0$. $Q = nC_V\,\Delta T = (0.325 \text{ mol})(20.85 \text{ J/mol} \cdot \text{K})(50 \text{ K}) = 339$ J.

(c) $e = \dfrac{W}{Q_H} = \dfrac{405 \text{ J} - 274 \text{ J}}{1422 \text{ J} + 339 \text{ J}} = 0.0744 = 7.44\%$.

$$e_{\text{Carnot}} = 1 - \frac{T_C}{T_H} = 1 - \frac{250 \text{ K}}{450 \text{ K}} = 0.444 = 44.4\%;$$

e_{Carnot} is much larger.

Reflect: $Q_{\text{tot}} = +1422 \text{ J} + (-1355 \text{ J}) + (-274 \text{ J}) + 339 \text{ J} = 132$ J. This is equal to W_{tot}, apart from a slight difference due to rounding. For a cycle, $W_{\text{tot}} = Q_{\text{tot}}$, since $\Delta U = 0$.

16.55. Set Up: The sun radiates heat energy at a rate $H = Ae\sigma T^4$. The rate at which the sun absorbs heat from the surrounding space is negligible, since space is so much colder. This heat flows out of the sun at 5800 K and into the surrounding space at 3 K. From Appendix F, the radius of the sun is 6.96×10^8 m. The surface area of a sphere with radius R is $A = 4\pi R^2$.

Solve: (a) In 1 s the quantity of heat radiated by the sun is

$$|Q| = Ae\sigma t T^4 = 4\pi R^2 e\sigma t T^4$$

$$|Q| = 4\pi (6.96 \times 10^8 \text{ m})^2 (1.0)(5.67 \times 10^{-8} \text{ W/m}^2 \cdot \text{K}^4)(1.0 \text{ s})(5800 \text{ K})^4 = 3.91 \times 10^{26} \text{ J}$$

$$\Delta S = \frac{-3.91 \times 10^{26} \text{ J}}{5800 \text{ K}} + \frac{+3.91 \times 10^{26} \text{ J}}{3 \text{ K}} = +1.30 \times 10^{26} \text{ J/K}$$

(b) The process of radiation is irreversible; this heat flows from the hot object (sun) to the cold object (space) and not in the reverse direction. This is consistent with the answer to part (a). We found $\Delta S_{\text{universe}} > 0$ and this is the case for an irreversible process.

Reflect: The entropy of the sun decreases because there is a net heat flow out of it. The entropy of space increases because there is a net heat flow into it. But the heat flow into space occurs at a lower temperature than the heat flow out of the sun and the net entropy change of the universe is positive.

16.57. Set Up: ab is at constant pressure, bc is at constant volume, and ca is isothermal. For H_2, $C_V = 20.42$ J/mol $\cdot$ K and $C_p = 28.74$ J/mol $\cdot$ K. For a constant pressure process, $Q = nC_p \Delta T$ and $n\,\Delta T = \dfrac{p\,\Delta V}{R}$, so $Q = \left(\dfrac{C_p}{R}\right)p\,\Delta V$. For a constant volume process, $Q = nC_V \Delta T$ and $n\,\Delta T = \dfrac{V\,\Delta p}{R}$, so $Q = \left(\dfrac{C_V}{R}\right)V\,\Delta p$. For an isothermal process, $\Delta U = 0$ and $Q = W = nRT\ln(V_2/V_1)$. 1 atm $= 1.013 \times 10^5$ Pa

Solve: (a) $pV = nRT$ so heat enters the gas when pV increases. In an isothermal process heat leaves the gas when V decreases. So, in this cycle heat enters in process ab and leaves in processes bc and ca.

(b) Q_H equals Q for process ab, so

$$Q_H = \left(\frac{C_p}{R}\right)p\,\Delta V = \left(\frac{28.74 \text{ J/mol} \cdot \text{K}}{8.315 \text{ J/mol} \cdot \text{K}}\right)(0.700 \text{ atm})(1.013 \times 10^5 \text{ Pa/atm})(0.070 \text{ m}^3) = 1.72 \times 10^4 \text{ J}$$

For ab, $W = p\,\Delta V = (0.700 \text{ atm})(1.013 \times 10^5 \text{ Pa/atm})(0.070 \text{ m}^3) = 4.96 \times 10^3$ J.
For bc, $W = 0$.
For ca, $W = nRT_a \ln(V_a/V_c)$.

$$T_a = \frac{p_a V_a}{nR} = \frac{(0.700 \text{ atm})(1.013 \times 10^5 \text{ Pa/atm})(0.0300 \text{ m}^3)}{(0.850 \text{ mol})(8.315 \text{ J/mol} \cdot \text{K})} = 301 \text{ K}$$

$$W = (0.850 \text{ mol})(8.315 \text{ J/mol} \cdot \text{K})(301 \text{ K})\ln\left(\frac{0.0300 \text{ m}^3}{0.100 \text{ m}^3}\right) = -2.56 \times 10^3 \text{ J}$$

The total work for the cycle is 4.96×10^3 J $+ (-2.56 \times 10^3$ J$) = 2.40 \times 10^3$ J.

$$e = \frac{W}{Q_H} = \frac{2.40 \times 10^3 \text{ J}}{1.72 \times 10^4 \text{ J}} = 0.140 = 14.0\%$$

Reflect: For a cycle, $\Delta U = 0$ and $Q_{\text{tot}} = W_{\text{tot}}$. The total heat flow must be 2.40×10^3 J, but in calculating e we use only the heat that flows into the engine.

16.59. Set Up: For a heat engine, $W = Q_C + Q_H$. $W > 0$, $Q_H > 0$ and $Q_C < 0$. $\Delta S = \dfrac{Q}{T}$. Q_H flows out of the hot reservoir at $T_H = 418$ K and Q_C flows into the cold reservoir at $T_C = 283$ K. The greatest efficiency possible between these two reservoirs is the Carnot efficiency.

Solve: (a) $W = Q_C + Q_H = -107.5$ J $+ 125.0$ J $= 17.5$ J

(b) $e = \dfrac{W}{Q_H} = \dfrac{17.5 \text{ J}}{125 \text{ J}} = 0.140 = 14.0\%$

(c) For the environment,

$$\Delta S = \frac{-Q_H}{T_H} + \frac{Q_C}{T_C} = \frac{-125.0 \text{ J}}{418 \text{ K}} + \frac{107.5 \text{ J}}{283 \text{ K}} = 0.0808 \text{ J/K}$$

(d) $e_{\text{Carnot}} = 1 - \dfrac{T_C}{T_H} = 1 - \dfrac{283 \text{ K}}{418 \text{ K}} = 0.323 = 32.3\%$

Reflect: The efficiency of this cycle is less than half the efficiency of a Carnot cycle operating between the same two hot and cold reservoirs.